USAF WARRIOR STUDIES

Richard H. Kohn and Joseph P. Harahan
General Editors

THE LITERATURE OF AERONAUTICS, ASTRONAUTICS, AND AIR POWER

Richard P. Hallion
Air Force Flight Test Center

OFFICE OF AIR FORCE HISTORY
UNITED STATES AIR FORCE
WASHINGTON, D.C. 1984

Library of Congress Cataloging in Publication Data

Hallion, Richard.
The Literature of Aeronautics, Astronautics, and Air Power.

(USAF warrior studies)
1. Aeronautics, Military—Bibliography. 2. Air power—Bibliography. 3. Astronautics, Military—Bibliography. I. United States. Air Force. Office of Air Force History. II. Title.
Z6724.A25H34 1984 (UG630) 016.3584 84-2322
ISBN 0-912799-14-5

Project Warrior Studies are published by the Office of Air Force History. The author is responsible for the scholarly judgments expressed herein, and his views should not be construed to be the official position of the United States Air Force or the Department of Defense.

For sale by the Superintendent of Documents, U.S. Government Printing Office
Washington, D.C. 20402

Foreword

The publication of *The Literature of Aeronautics, Astronautics, and Air Power* is part of a continuing series of historical studies from the Office of Air Force History in support of Project Warrior.

Project Warrior seeks to create and maintain within the Air Force an environment where Air Force people at all levels can learn from the past and apply the warfighting experiences of past generations to the present. When General Lew Allen, Jr. initiated this project in 1982, he called for the "continuing study of military history, combat leadership, the principles of war and, particularly, the applications of air power." All of us in the Air Force community can benefit from such study and reflection. The challenges of today and the future demand no less.

CHARLES A. GABRIEL, General, USAF
Chief of Staff

The Author

RICHARD P. HALLION is the Center Historian at the Air Force Flight Test Center, Edwards AFB, California. He received his Ph.D. in History from the University of Maryland in 1975. From 1974-1980, he served as the Curator of Science and Technology and subsequently as Curator of Space Science and Exploration at the National Air and Space Museum, Smithsonian Institution, Washington, D.C. In those same years he was an Associate Professor of History at the University of Maryland and a Visiting Fellow at Yale University. Dr. Hallion is the author of numerous works in aerospace history, including *Supersonic Flight: Breaking the Sound Barrier and Beyond* (New York: Macmillan Co, 1972); *Legacy of Flight: The Guggenheim Contribution to American Aviation* (Seattle: University of Washington Press, 1977); *Test Pilots: The Frontiersmen of Flight* (Garden City, N.Y.: Doubleday & Co, 1981); *Rise of the Fighters: Air Combat in World War I* (Annapolis: Nautical and Aviation Publishing Co, 1984); *On the Frontier: Flight Research at Dryden, 1946-1981* (Washington: NASA, 1984). In addition, he edited *The Wright Brothers: Heirs of Prometheus* (Washington: Smithsonian Institution Press, 1978), and with Tom Crouch, *Apollo: Ten Years Since Tranquillity Base* (Washington: Smithsonian Institution Press, 1979).

United States Air Force
Historical Advisory Committee
(As of October 1, 1984)

Contents

Introduction

Since its inception in 1982, Project Warrior has revived interest inside the Air Force in using history to sharpen professional knowledge of air power. Project Warrior has also expanded interest in the heritage of the U.S. Air Force in peace and in war. The bibliographical essay published here as *The Literature of Aeronautics, Astronautics, and Air Power* is meant to provide readers with a guide to the vast collections of books and articles available today in libraries and from publishers.

It was written originally in early 1982 by Dr. Richard P. Hallion as a Project Warrior monograph at the Air Force Systems Command's Edwards Flight Test Center, California. Drawing on his experience as an author, a curator at the Smithsonian Institution's National Air and Space Museum, and as a history professor at the University of Maryland, Dr. Hallion recognized the need of scientists and engineers for a guide to the research and technical literature on the history of aeronautics and astronautics. Publication here of a revised, reorganized, and expanded version reflects our belief that what was so valuable to the flight test community will, in different form, be of similar use to the rest of the Air Force, to civilian scholars, and to aviation enthusiasts generally.

Dr. Hallion's essay, while revised in size, scope, and emphasis, retains the flavor of its original purpose and reflects the author's interests, background, and professional judgments. In part, the essay also reflects suggestions made by the editors, especially on the growth and development of air power in the twentieth century.

The publishing of this work in book form was made possible by several individuals in the Office of Air Force History. Joseph P. Harahan served as general manuscript editor, checking virtually every citation in the text. Richard H. Kohn interweaved some titles on air power from an earlier bibliographic effort of his own. Herman Wolk and Colonel John F. Shiner, USAF, critiqued the essay and provided excellent advice at critical points.

R. H. K.
J. P. H.

THE LITERATURE OF AERONAUTICS, ASTRONAUTICS, AND AIR POWER

General Reference Sources

There are relatively few works that explicitly treat the history of aviation with a view to tracing how the craft of aerospace technology emerged. The history of technology within aviation history is an area that is, by and large, wide-open for serious historical investigation. Currently, there is no one single overall survey of the history of aeronautics and astronautics that can be considered definitive. Roger E. Bilstein, *Flight in America, 1900–1983: From the Wright Brothers to the Astronauts* (Baltimore: Johns Hopkins University Press, 1984) represents one effort to fill this need. The best reference is a series of essays edited by Eugene M. Emme, *Two Hundred Years of Flight in America: A Bicentennial Survey* (San Diego, Calif.: American Astronautical Society, 1977). This book is very useful both as an introductory reader and as a text in aerospace history courses. John D. Anderson, *Introduction to Flight: Its Engineering and History* (New York: McGraw-Hill, 1978) is a noteworthy blending of introductory engineering theory and historical interpretation. Ronald Miller and David Sawers, *The Technical Development of Modern Aviation* (New York: Praeger, 1970) is a good reference on the development of air transport technology, though less satisfactory on the high technology of advanced military systems. J. L. Nayler and Ernest Ower, *Aviation: Its*

Technical Development (London: Peter Owen, Vision Press, 1965) is a generally useful introduction to the history of aerospace technology. Oliver Stewart, *Aviation: The Creative Ideas* (New York: Praeger, 1966) is a thoughtful and well-written series of essays on one man's perspective on the history of aviation, but readers are cautioned that his chapter on the Wrights and Clement Ader (a French pioneer) is dangerously misleading. Ader did not, as Stewart claims, understand aeronautical concepts anywhere near the depth of the Wrights. Richard P. Hallion, "The Rise of Air and Space," in *Astronautics and Aeronautics,* 19, No. 5, (May 1981) furnishes an introductory overview to the general history of aeronautics and astronautics.

The field of aerodynamics is one that requires a good historical survey bringing it up to date. Theodore von Karman's, *Aerodynamics* (New York: McGraw-Hill, 1963), a collection of six lectures he presented at Cornell University in 1953, is still the best introduction to the history of aerodynamics. A highly technical and exhaustive survey of aerodynamics history through 1930 is an essay by R. Giacomelli and E. Pistolesi in William F. Durand's series, *Aerodynamic Theory: A General Review of Progress,* Vol 1, (New York: Dover Publications, 1963). This series, incidentally, has itself taken on major importance to the understanding of the evolution of aerospace technology. The Dayton-Cincinnati Section of the American Institute of Aeronautics and Astronautics (the national professional organization of aerospace engineers, scientists, and technologists) has issued two compilations of papers relating to the technical development of aviation. Though the papers vary in quality, the works are very useful as references: Jay D. Pinson, compiler, *Diamond Jubilee of Powered Flight: The Evolution of Aircraft Design* (Dayton: AIAA, 1978), and *The Evolution of Aircraft Wing Design* (Dayton: AIAA, 1980). Both works resulted from historical symposia.

The International Academy of Astronautics of the International Astronautical Federation has sponsored an annual historical symposium since 1967; the papers presented at these symposia have been from a truly international body of pioneers and distinguished historians, and, as such, are an invaluable compilation of source material on the early history of astronautics. The 1967–1972, inclusive, symposia proceedings have been published in the volumes Frederick C. Durant and George S. James, eds, *First Steps Towards Space* (Washington: Smithsonian Institution Press, 1974), and R. Cargill Hall, ed, *Essays on the History of Rocketry and Astronautics: Proceedings of the Third Through the Sixth History Symposia of the International Academy of Astronautics*, Vols I, II, (Washington: NASA, 1977). Future volumes will treat subsequent symposia.

The industrial and "think tank" perspectives on aeronautical and astronautical development are ones that have traditionally been slighted. Two uncritical but nevertheless valuable looks at aerospace from the corporate and laboratory environment are William A. Schoneberger, *et al, Seven Decades of Progress: A Heritage of Aircraft Turbine Technology* (Fallbrook, Calif.: Aero Publishers, 1979), which examines the gas turbine work of General Electric, and Everett T. Welmers, *et al, The Aerospace Corporation—Its Work: 1960–1980* (Los Angeles, Calif.: Aerospace Corporation, 1980), which is a history of the role one major American aerospace think tank has played in recent aerospace development.

One of the most significant international organizations in the development of aviation technology was the National Advisory Committee for Aeronautics (1915–1958), the predecessor of the present-day National Aeronautics and Space Administration. The NACA and its operations are rich subjects for historical research, although an administrative and organizational study by Professor Alex Roland of Duke University is forthcoming. George W. Gray, *Frontiers of Flight: The Story of NACA Research* (New York: Knopf, 1948) was an early attempt to discuss the work of the NACA, and, though flawed by a generally uncritical tone, the book is very valuable. Frank W. Anderson, *Orders of Magnitude: A History of NACA and NASA, 1915–1980* (Washington: NASA, 1981) provides a good introductory look at the two agencies. Jerome Hunsaker, "Forty Years of Aeronautical Research," in the *Smithsonian Report for 1955* (Washington: Smithsonian Institution Press, 1956) gives a good overall survey of the agency's work during the greater part of its existence by one of its last chairmen. John Becker, *The High-Speed Frontier: Case Histories of Four NACA Programs, 1920–1950* (Washington: NASA, 1980) is an excellent engineering memoir and critical analysis by a former senior NACA–NASA engineer and administrator. Donald D. Baals and William R. Corliss, *Wind Tunnels of NASA* (Washington: NASA, 1981) is an excellent survey of the development of the wind tunnel in the United States and its contributions to aerospace technology. David A. Anderton, *Sixty Years of Aeronautical Research: 1917–1977* (Washington: NASA, 1978) is a good introduction to the research activities of NASA's Langley Research Center, the first and most important of the aeronautical laboratories.

There have been many attempts to collect and publish essays and articles dealing with aviation. Of these various collections, four deserve special mention: Eugene M. Emme, *The Impact of Air Power: National Security and World Politics* (Princeton, N.J.: D. Van Nostrand, 1959) is an excellent compilation of source documents and articles; John F. Loosbrock and Richard M. Skinner, *The Wild Blue: The Story of American Airpower* (New York: G. P. Putnam's Sons, 1961) is an excellent and informative anthology tracing the development of flight; Joseph B. Roberts and Paul L. Briand, eds, *The Sound of Wings: Reading for the Air Age* (New York: Henry Holt and Co, 1957) includes a number of essays and works that relate aviation to social and cultural perspectives; and finally, Alfred F. Hurley and Robert C. Ehrhart, eds, *Air Power and Warfare: The Proceedings of the 8th Military History Symposium, United States Air Force Academy, 18–20 October 1978* (Washington: Office of Air Force History and USAF Academy, 1979) contains excellent recent scholarship and analysis by both scholars and airpower participants.

Chronology can be said to be the framework of history. Fortunately for those interested in the history of aviation there have already been some well executed chronologies tracing the development of flight. A very useful introductory chronology can be found in the first volume of Michael J. H. Taylor, *et al*, *Jane's Encyclopedia of Aviation*, 5 vols (London: Jane's

Publishing Company, 1980), though, it must be noted, neither the chronology nor the subjects covered are treated as exhaustively as the title of this five-volume series suggests. A truly significant chronological effort has been that of the NASA History Office, with a series of NASA-sponsored publications that have presented a detailed, world-wide chronology of flight ranging from 1915 through 1976. The first of these is Eugene M. Emme's *Aeronautics and Astronautics: An American Chronology of Science and Technology in the Exploration of Space, 1915–1960* (Washington: NASA, 1961). Two subsequent chronologies, *Aeronautical and Astronautical Events of 1961* (Washington: Government Printing Office, 1962), and *Aeronautical and Astronautical Events of 1962* (Washington: Government Printing Office, 1963), were published as NASA reports to the U.S. House of Representatives' Committee on Science and Astronautics. Since then, the NASA History Office has issued an annual *Astronautics and Aeronautics* (Washington: NASA, 1964–present) volume, which includes a detailed, annotated international chronology of events relevant to the history of aerospace. Welman A. Shrader, *Fifty Years of Flight: A Chronicle of the Aviation Industry in America, 1903–1953* (Cleveland: Eaton, 1953), while dated, is an excellent reference. L. G. S. Payne, *Air Dates* (New York: Praeger, 1957) treats European, especially British, aviation.

Roger Pineau's exhibit catalog, *Ballooning, 1782–1972* (Washington: Smithsonian Institution Press, 1972), prepared for the opening of an exhibition gallery at the Smithsonian Institution, contains an excellent chronology of lighter-than-air flight. Arnold E. Briddon, *et al, FAA Historical Fact Book: A Chronology, 1926–1971* (Washington: Department of Transportation, 1974) is a reliable and very complete chronology of matters relating to American civil aviation. Clarke Van Vleet and William J. Armstrong, *United States Naval Aviation, 1910–1980* (Washington: Naval Air Systems Command, 1981) have written the definitive chronology on American naval aviation. The U.S. Air Force, *A Chronology of American Aerospace Events* (Washington: Office of Information Services, 1959) is a good source of early Air Force and other American aeronautical information. Supplements have been issued periodically by the Air Force Office of Public Affairs, Washington, D.C. Kit C. Carter and Robert Mueller, *The*

Army Air Forces in World War II: Combat Chronology, 1941–1945 (Washington: Office of Air Force History, 1973) is, as the title implies, an exhaustive chronological accounting of the day-to-day activities of the AAF at war. Finally, the American Institute of Aeronautics and Astronautics has supported an extensive chronological effort by Frank H. Winter of the National Air and Space Museum, in conjunction with Richard P. Hallion and Frank Robert Van Der Linden. Each month, in the AIAA journal *Astronautics and Aeronautics,* the AIAA publishes a column entitled "Out of the Past," treating events that occurred in aviation and aerospace 25, 50, and 100 years previously. This column has been appearing regularly since 1972, and is especially valuable because it references its entries to applicable sources.

Oddly, there have been very few biographical guides to the individuals who have made aviation what it is today. Howard S. Wolko, *In the Cause of Flight: Technologists of Aeronautics and Astronautics* (Washington: Smithsonian Institution Press, 1981) is an excellent reference, with short biographical sketches and lengthy and incisive surveys of the various fields within aviation technology, together with their "players." Shirley Thomas, *Men of Space,* 8 vols (Philadelphia: Chilton, 1960–) is a useful guide to the pioneers of the space age, written in a breezy, anecdotal style. Aside from these, however, one must look to contemporary reference sources and to the few published biographies of specific individuals cited earlier. G. Edward Pendray, *The Guggenheim Medalists: Architects of the Age of Flight* (New York: The Guggenheim Medal Board of Award of the United Engineering Trustees, Inc, 1964) is a handy reference to some of the key technologists, scientists, practitioners, and managers of aviation's early days, who received the prestigious Daniel Guggenheim Medal for great achievement in the cause of flight. Three "Who's Who" books are especially good references: Lester D. Gardner, *Who's Who in American Aeronautics* (Los Angeles: Floyd Clymer Publications, 1925, reissued, 1974), also called *The Blue Book of American Airmen;* the Writers' Program of the Work Projects Administration, *Who's Who in Aviation, 1942–43* (New York: Ziff-Davis Publishing Co, 1942); and Wayne W. Parrish, *Who's Who in World Aviation, 1955* (Washington: American Aviation Publication, 1955). The National Aeronautical Institute, *Who's Who in Aviation and Aerospace* (Boston: National Aeronautical Institute, 1983) is a voluminous recent guide to individuals working in the field, many of whom started their careers in the mid-to-late 1940s. Finally, one useful reference that can contribute to knowledge of distinguished individuals who worked within the aerospace profession is by the National Air and Space Museum Library, *International Handbook of Aerospace Awards and Trophies* (Washington: Smithsonian Institution Press, 1978), prepared under the direction of Catherine D. Scott.

So-called "coffee table" books often can be a surprisingly good source of both information and visual references. They must be used with care, however, for a number exist that are misleading or incorrect. Three are considered especially reliable and useful: Charles H. Gibbs-Smith, *Flight Through the Ages* (New York: Crowell, 1974) combines chronology with excellent illustrations to provide the reader with a thorough and accurate account of aviation history; John W. R. Taylor and Kenneth Munson, *History of Aviation* (New York: Putnam, 1978); and *The Lore of Flight* (Gothenburg, Sweden: Tre Tryckare Cagner & Co, 1970), which is an excellent introduction into the functional uses of aircraft and spacecraft, as well as their history. As mentioned earlier, there still is no single reference or group of references that, in a concise form, furnishes a useful overview of the history of aeronautics and astronautics. Time-Life Books' *Epic of Flight* series, cited where appropriate throughout this essay, constitutes a good attempt at such a work, but still does not address post-1945 aviation in detail.

The actual flight vehicles of aerospace history have held a fascination for writers for years, and the literature on individual airplanes is so voluminous as to actually clutter the field. *Jane's All the World's Aircraft* (London: Jane's Publishing Co, 1909–present) has long been the standard yearly reference on aircraft development; some issues, such as its 1919 and 1945 issues, are classic references. Later (post-1945) editions of Jane's offer considerably more detail than earlier ones. Two other Jane's publications, one by Horace F. King and John W. R. Taylor, *Jane's 100 Significant Aircraft, 1909–1969* (London: Jane's Publishing Co, 1969), and the previously mentioned five-volume *Jane's Encyclopedia of Aviation* are significant. The latter work, despite its title, is really a guide to the world's air forces, airlines, and the various aircraft built by nations through the years. It is, unfortunately, flawed by minor errors of fact and interpretation, and thus must be used with caution. John W. R. Taylor, *Combat Aircraft of the World* (New York: G. P. Putnam's Sons, 1969) is a comprehensive introduction to the combat aircraft produced from 1909 through the late 1960s, and is very useful. Ray Wagner, *American Combat Planes,* 3rd revised edition (Garden City, N.Y.: Doubleday, 1982) is an excellent introduction to the American perspective of military aircraft development. An authoritative reference work is F. Gordon Swanborough and Peter M. Bowers, *United States Military Aircraft Since 1909* (London: Putnam, 1971). Claudia Oakes, *Aircraft of the National Air and Space Museum* (Washington: Smithsonian Institution Press, 1981) is an excellent guide to the holdings of the Smithsonian Institution, one of the world's great aircraft collections, and contains excellent aircraft "biographies" written by the NASM's curatorial staff. A special mention must be made of the *Putnam Aeronautical Books* series published since the early 1960s by Putnam & Company, Ltd, 9 Bow Street, London, England. This series consists of individual volumes, often running to many hundreds of pages on aircraft developed by specific manufacturers, such as McDonnell-Douglas and Bristol. The books are heavily illustrated with photographs and drawings, and contain much useful information on the growth of the world aircraft industry. The *Profile Publication* series (London: Hills and Lacy, 1965–1975), a collection of over 250 heavily illustrated pamphlets on individual aircraft types (such as the Fokker D–VII or the Douglas DC–3), is an excellent detailed source of information on particular aircraft. Each pamphlet was written by an authority on the particular aircraft. This series was issued in several volumes in the United States by Doubleday & Co, and, though now long out of print, is still available as a library reference. Finally, Bill Gunston's *The World's Greatest Airplanes: The Story of the Men Who Built Them and How They Came to Be* (New York: Elsevier-Dutton Publishing Co, 1980) is a good popular introduction told with style and with a goodly number of anecdotes about the world's great aircraft companies, and how they evolved.

A number of specialized guides to research in aerospace history have been produced through the years, in addition to such well-known general guides as the *Reader's Guide to Periodical Literature*. The following bibliographical sources are all recommended for their completeness and excellence: Samuel Duncan Miller, compiler, *An Aerospace Bibliography* (Washington: Office of Air Force History, 1978); Mary Ann Cresswell and Carl Berger, compilers, *United States Air Force History: An Annotated Bibliography* (Washington: Office of Air Force History, 1971); Katherine Murphy Dickson, compiler, *History of Aeronautics and Astronautics: A Preliminary Bibliography* (Washington: NASA, 1968); Alex Roland, *A Guide to Research in NASA History* (Washington: NASA, 1983); Jacob Neufeld, compiler, *United States Air Force History: A Guide to Monographic Literature, 1943–1974* (Washington: Office of Air Force History, 1977); Lawrence J. Paszek, compiler, *United States Air Force History: A Guide to Documentary Sources* (Washington: Office of Air Force History, 1973), the latter containing an excellent guide to the holdings of the Library of Congress, the National Archives, and the USAF Historical Research Center; and, finally, John J. Looney, compiler, *Bibliography of Space Books and Articles from non-Aerospace Journals, 1957–1977* (Washington: NASA, 1979).

The Prehistory of Flight, Antiquity to 1783

Practical aviation as it is known today began with the first balloon flights of the Montgolfier brothers and J. A. C. Charles in the year 1783. The social impact of these flights was considerable; it was reflected in art, literature, and interior design. "Balloonmania" resulted in references to 1783 as the "Year of Miracles," and the appearance of balloon prints, chandeliers, chairs, and the like. For the first time, humanity was able to view the earth from the air. However, the accomplishment of practical flight in 1783 must not disguise the fact that flight had been one of the strongest of human aspirations, dating to antiquity and ancient myths. The single best reference work on early precursors of flight is Clive Hart, *The Dream of Flight: Aeronautics from Classical Times to the Renaissance* (New York: Winchester Press, 1972), a remarkable compilation of material including extensive citations from primary source documents. A standard reference on the history of aviation is Charles Gibbs-Smith, *The Aeroplane: An Historical Survey of Its Origins and Development* (London: Her Majesty's Stationary Office, 1960), and his later revised edition, *Aviation: An Historical Survey from its Origins to the End of World War II* (London: Her Majesty's Stationary Office, 1970). Until his death in 1981, Gibbs-Smith was the doyen of aeronautical historians and the recognized authority on aviation before the Wright's first flight. Two intriguing studies of specific uses of aeronautics and attempts to fly in medieval times are Lynn White's "Eilmer of Malmesbury: An Eleventh Century Aviator," *Technology and Culture,* 19, No. 2 (Spring 1961) and "Medieval Uses of Air," *Scientific American,* 223, No. 12 (August, 1970). The history of ballooning has not received the attention that it's due. Two new works, however, Tom D. Crouch's, *The Eagle Aloft: Two Centuries of the Balloon in America* (Washington: Smithsonian Institution Press, 1983), and Charles S. Gillispie's, *The Montgolfier Brothers* (Princeton: Princeton University Press, 1983), will fill this gap. Two other references are: *The Romance of Ballooning: The Story of the Early Aeronauts* (New York: The Viking Press, 1971), and Donald Dale Jackson, *The Aeronauts* (Alexandria, Va.: Time-Life Books, 1980).

The Era of Discovery, 1783–1903

During the nineteenth century, a technological base was established that enabled the development of the first heavier-than-air flying machines. This work began with the theoretical and practical experiments and studies of Sir George Cayley, continued through such individuals as Stringfellow and Henson, and concluded with the epochal work of Otto Lilienthal and the Wrights. Along the way, an interesting mix of successful and unsuccessful practitioners advanced the understanding of what was needed to produce actual airplanes. Again, much of the most significant historical writing in this field has come from the pen of Charles Gibbs-Smith, notably his *The Invention of the Aeroplane, 1799–1909* (London: Faber & Faber, 1965), and his encyclopedic *A Directory and Nomenclature of the First Aeroplanes, 1809–1909* (London: Her Majesty's Stationary Office, 1966), a particularly useful illustrated guide to the aircraft and attempted aircraft of that period. Tom D. Crouch, *A Dream of Wings: Americans and the Airplane, 1875–1905* (New York: W. W. Norton & Co, 1981) thoroughly examines the American aeronautical community, with insightful analysis of such figures as Octave Chanute, Samuel Langley, and the Wright brothers. Chanute's own *Progress in Flying Machines* (New York: Forney, 1894) remains a valuable reference to the climate of aeronautical opinion in the 1890s, as does Jeremiah Milbank, *The First Century of Flight in America* (Princeton: Princeton University Press, 1943). The latter was the first scholarly examination of the early history of aviation and flight attempts in the United States. A recent work by Joseph J. Corn, *The Winged Gospel: America's Romance with Aviation, 1900–1950* (New York: Oxford University Press, 1983) provides a social history of aviation.

Marvin McFarland's impressive two-volume *The Papers of Wilbur and Orville Wright* (New York: McGraw-Hill Book Co, 1953) constitutes an indispensible reference on the thoughts and activities of the two brothers from Dayton, Ohio. Fred Kelly's excellent though dated biography, *The Wright Brothers* (New York: Harcourt, Brace, 1943) has now been supplanted by Harry Combs' masterful *Kill Devil Hill: Discovering the Secret of the Wright Brothers* (Boston: Houghton Mifflin, 1979). Combs' work goes far in demythologizing the Wrights and correctly portraying them as intuitive engineers who carefully plotted and organized their progression from the drawing board to the sands of Kitty Hawk. Richard Hallion, ed, *The Wright Brothers: Heirs of Prometheus* (Washington: Smithsonian Institution Press, 1978) is a collection of essays from experts who have studied the work of the Wrights and its impact upon society. This volume also contains Orville Wright's own account of the brothers' preparations for flight, and is a useful introduction to the Wrights and their accomplishments. A colorful and reliable history of efforts leading to the Wrights' triumph at Kitty Hawk is by Valerie Moolman, *The Road to Kitty Hawk* (Alexandria, Va.: Time-Life Books, 1980). This book, part of the Time-Life *Epic of Flight* series, is, like most of the works in the series, replete with numerous photographs and drawings and a heavily anecdotal text, being particularly suited to those needing a readable and popular introduction to the subject.

The Development of Practical Airplanes, 1903–1918

The time period running from the Wrights through the First World War was a particularly challenging and fruitful time in aeronautics. Critical advances were made in aerodynamics, propulsion, structures, and controls technology, enabling a four-fold increase in aircraft flight speeds. By the end of the First World War, aircraft design technology had advanced to the point where operators could confidently plan long-range flights across the North Atlantic and through Southeast Asia to Australia. The airplane had proven itself a major military weapon, forcing military planners to rethink strategy and tactics in light of its capabilities. Charles Gibbs-Smith, *The Rebirth of European Aviation, 1902–1908* (London: Her Majesty's Stationary Office, 1974) offers a clear analysis of the Wrights' impact upon a European aeronautical community suddenly shocked into awareness of its technological inferiority. Several useful popular accounts of aviation's pioneering days exist, notably, Walter T. Bonney, *The Heritage of Kitty Hawk* (New York: Norton, 1962); Henry Serrano Villard, *Contact: The Story of the Early Birds* (New York: Crowell, 1968); and Sherwood Harris, *The First to Fly: Aviation's Pioneer Days* (New York: Simon & Schuster, 1970). Sir Geoffrey de Havilland's reissued autobiography, *Sky Fever* (Shrewsbury, England: Airlife Publications, 1979), offers an often amusing and informative look at the trials and tribulations of an early British aircraft builder and pilot. Constance Babington-Smith's *Testing Time: The Story of*

British Test Pilots and Their Aircraft (New York: Harper & Brothers, 1961), is an excellent reference to the pioneer days of flight in Great Britain. Harald Penrose, a noted British test pilot, has written a good, if controversial, history tracing pre-Great War aviation in Great Britain, *British Aviation: The Pioneer Years, 1903–1914* (London: Putnam, 1967). Cecil R. Roseberry, *Glenn Curtiss: Pioneer of Flight* (Garden City, N.Y.: Doubleday & Co, 1972) is the standard biography of this important American pioneer, who was the Wrights' chief rival in the early days. Louis S. Casey's recent *Curtiss: The Hammondsport Era, 1907–1915* (New York: Crown Publishers, 1981) and Peter M. Bowers, *Curtiss Aircraft: 1907–1947* (London: Putnam, 1979) offer surprisingly detailed and accurate information on the aircraft themselves.

Owen S. Lieberg, *The First Air Race: The International Competition at Reims, 1909* (Garden City, N.Y.: Doubleday & Co, 1974) is a popularly written and informative account of the first great air meet, enabling critical comparisons of technological approaches to be made. Tom D. Crouch, *Bleriot XI: The Story of a Classic Aircraft* (Washington: Smithsonian Institution Press, 1982) is a valuable account of the work of French pioneer Louis Bleriot, developer of one of early aviation's most significant flying vehicles. Interestingly, Allen Wheeler, *Building Aeroplanes for "Those Magnificent Men"* (London: G. T. Foulis & Co, 1965) wrote an account of constructing flying replicas for the film *Those Magnificent Men in Their Flying Machines*. It is one of the best sources of information on the design, construction techniques, flying, and handling qualities of early airplanes. The actual pioneers of the early days have not received the biographical treatment that is their due, though Claudia Oakes, *United States Women in Aviation Through World War I* (Washington: Smithsonian Institution Press, 1978) and Howard S. Wolko, *In the Cause of Flight: Technologists of Aeronautics and Astronautics*, cited previously, are useful steps towards remedying this situation. Overall, Curtis Prendegast, *The First Aviators* (Alexandria, Va.: Time-Life Books, 1980), part of the Time-Life series, is a reliable and well-written popular introduction to the people of the period.

The First World War in the air has been the subject of so many works that it is virtually impossible to discuss all of them in a coherent essay.

LITERATURE

Military development before the war was marked by lethargy inspired by official military reluctance to sanction creation of air arms. American developments are covered in Juliette A. Hennessey, *The United States Army Air Arm, April 1861 to April 1917*, USAF Historical Studies, No. 98 (Maxwell AFB, Ala: Air University, 1958). Charles DeForest Chandler and Frank P. Lahm, *How Our Army Grew Wings: Airmen and Aircraft Before 1914* (New York: The Ronald Press Company, 1943) is part history, part memoir by two pioneers in American military aviation. Once war broke out, however, expansion was rapid if often haphazard. Again, a good popular reference is Ezra Bowen, *Knights of the Air* (Alexandria, Va.: Time-Life Books, 1980).

Readers desiring more comprehensive and in-depth coverage are encouraged to examine the following works. George van Deurs, a pioneering Navy pilot, has examined that service's introduction to the airplane in *Wings for the Fleet: A Narrative of Naval Aviation's Early Development, 1910–1916* (Annapolis, Md.: U.S. Naval Institute Press, 1966). Peter B. Mersky, *U.S. Marine Corps Aviation 1912 to the Present* (Annapolis: Nautical and Aviation Publishing Company of America, 1983) details the long sweep of Marine Corps aviation. Henry "Hap" Arnold's autobiographical *Global Mission* (New York: Harper & Row, 1949) offers comparable insight into the early years of U.S. Army Aviation; for more information, readers should consult Alfred Goldberg, *et al, A History of the United States Air Force, 1907–1957* (Princeton: D. Van Nostrand Co, 1957). The British six-volume official history by Walter Raleigh and Henry A. Jones, *The War in the Air*, 6 vols (London: Oxford University Press and Clarendon Press, 1922-37) offers an in-depth look at the air war over the Western Front from an Allied perspective. Andrew Boyle, *Trenchard: Man of Vision* (London: Collins, 1962) is a penetrating and sympathetic study of a key advocate of airpower. Two "before and after" studies on early attempts to undertake strategic air operations by Imperial Germany are Douglas Robinson, *The Zeppelin in Combat: A History of the German Naval Airship Division, 1912–1918* (Seattle: University of Washington Press, 1980) and Raymond H. Fredette, *The Sky on Fire: The First Battle of Britain, 1917–1918, and the Birth of the Royal Air Force* (New York: Holt, Rinehart

& Winston, 1966), paperback edition (New York: Harcourt Brace and Jovanovich, 1976). An excellent overview is by John H. Morrow, *German Airpower in World War I* (Lincoln: University of Nebraska Press, 1982).

America's wartime aeronautical effort was marked by political scandal and mismanagement at home and dedicated professionalism and heroism abroad. Definitive studies of the home front and combat front are by I. B. Holley, *Ideas and Weapons: Exploitation of the Aerial Weapon by the United States during World War I: A Study in the Relationship of Technological Advance, Military Doctrine, and the Development of Weapons* (New Haven: Yale University Press, 1953), reprinted (Washington: Office of Air Force History, 1984), and James J. Hudson, *Hostile Skies: A Combat History of the American Air Service in World War I* (Syracuse: Syracuse University, 1968). Essential documents can be found in Maurer Maurer, ed, *The U.S. Air Service in World War I,* 4 vols.(Washington: Office of Air Force History, 1978–1979). A new work by Richard P. Hallion, *Rise of the Fighters: Air Combat in World War I* (Annapolis: Nautical and Aviation Publishing Co, 1984) examines the development of combat aircraft from a new perspective. Pilot accounts from all sides abound. Among the best are: Cecil Lewis, *Sagittarius Rising* (Harrisburg: Stackpole Books, 1963); Ernst Udet, *Ace of the Iron Cross* (Garden City, N.Y.: Doubleday, 1970); and Eddie Rickenbacker, *Fighting the Flying Circus* (Garden City, N.Y.: Doubleday, 1965). Denis Winter, *The First of the Few: Fighter Pilots of the First World War* (Athens: University of Georgia Press, 1983) uses mostly British sources to provide a penetrating analysis of the pilots and air-to-air combat experiences of the war. Works dealing with the aircraft of the First World War are almost too numerous to mention. Kenneth Munson, *Aircraft of World War I* (Garden City, N.Y.: Doubleday, 1968) is a reliable and well illustrated introductory guide to the frail machines that fought between 1914–1918. John Cuneo's impressive two-volume *Winged Mars* (Harrisburg: Military Service Publishing Co, 1945) is a thorough and incisive examination of the emergence of a military aviation consciousness in the opening years of the First World War. Overall, the best survey histories of the rise of military air power are by Robin Higham, *Air Power: A Concise History* (New York: St. Martin's Press, 1972) and Basil Collier, *A History of Air Power* (New York: Macmillan, 1974).

LITERATURE

The Rise of Air Transportation and Professionalization, 1918–1936

The interwar years through the mid-1930s were a particularly critical time for aeronautics. It was in this decade that the first practical air transport aircraft were developed, leading to the emergence of widespread air passenger and freight service. Also, military airpower doctrine was refined, and the roles and missions of military aircraft types more closely defined. The basic technology of flight underwent a revolution, with the development of powerful piston engines, efficient wing shapes, a variety of specialized devices for improved aerodynamic performance, the emergence of the monoplane (one-wing) configuration, and the appearance of the all-metal airplane. The profession of aerospace engineer also underwent a significant upgrading in skill level and training with the appearance of schools and departments of aerospace engineering, usually drawing upon the field of mechanical engineering and fluid mechanics for their technical and scientific background. An excellent introduction to this period can be found in Roger E. Bilstein, *Flight Patterns: Trends in American Aviation 1918–1929* (Athens, Ga.: University of Georgia Press, 1983).

By the beginning of 1919, the technology of aircraft and airships was advanced enough to permit the undertaking of long distance transoceanic flights. A good introduction to long distance aviation is David Nevin, *The Pathfinders* (Alexandria, Va.: Time-Life Books, 1980) which covers the pioneering efforts of such individuals as Alcock and Brown, Lindbergh, Wiley Post, and Howard Hughes. This excellent book is complemented by six others in the Time-Life series that also treat important themes of the interwar years: Douglas Botting, *The Giant Airships* (Alexandria, Va.: Time-Life Books, 1980); Paul O'Neil, *Barnstormers and Speed Kings* (Alexandria, Va.: Time-Life Books, 1981); Oliver E. Allen, *The Airline Builders* (Alexandria, Va.: Time-Life Books, 1981); David Nevin, *Architects of Air Power* (Alexandria, Va.: Time-Life Books, 1981); and Donald Dale Jackson, *Flying the Mail* (Alexandria, Va.: Time-Life Books, 1982). All are written in anecdotal fashion, and profusely illustrated. The U.S.

Navy's pioneering transatlantic flight of 1919 using the NC–4 flying boat is treated by Richard K. Smith, *First Across* (Annapolis: U.S. Naval Institute Press, 1973). The epochal transatlantic, transpacific, and arctic flights of the 1920s are the subject of a large body of literature. Of special value are: Joseph Hamlen, *Flight Fever* (Garden City, N.Y.: Doubleday & Co, 1971); Edward Jablonski, *Atlantic Fever* (New York: The Macmillan Co, 1972); Jack Huttig, *1927: Summer of Eagles* (Chicago: Nelson-Hall, 1979); and John Grierson, *Challenge to the Poles: Highlights of Arctic and Antarctic Aviation* (London: G. T. Foulis & Co, 1964), the definitive work by a well-known pioneering British pilot.

Charles A. Lindbergh and his solo flight across the North Atlantic in May 1927 inspired countless journalists and writers to write about the lanky airman. Unfortunately, most have written in a sensationalist fashion, ignoring the careful and methodical planning that went into the flight. Tom D. Crouch, ed, *Charles A. Lindbergh: An American Life* (Washington: Smithsonian Institution Press, 1977), a series of essays by experts in the field, has an excellent introduction to Lindbergh, his accomplishments, and the political controversy that surrounded him later in life. Lindbergh has been the subject of three general biographies: Kenneth Davis, *The Hero: Charles A. Lindbergh and the American Dream* (Garden City, N.Y.: Doubleday & Co, 1959); Walter S. Ross, *The Last Hero: Charles A. Lindbergh* (New York: Harper & Row, 1968); and Leonard Mosley, *Lindbergh: A Biography* (Garden City, N.Y.: Doubleday & Co, 1976). Davis and Ross are still the best; Lindbergh preferred the Ross work. Additionally, Wayne S. Cole, *Charles A. Lindbergh and the Battle Against American Intervention in World War II* (New York: Harcourt Brace Jovanovich, 1974) is the most reliable and authoritative study on Lindbergh's controversial political stand in the years before Pearl Harbor. Lindbergh himself was the author of numerous books and articles during his life, and these offer a rich treasury of materials on his life and wide breadth of interests. Especially recommended are: *We* (New York: Putnam, 1927), his account of the Paris flight written just after his return; *The Spirit of St. Louis* (New York: Scribner, 1953), and *Autobiography of Values* (New York: Harcourt Brace Jovanovich, 1978), the latter published following his death in 1974. Donald E. Keyhoe's *Flying with Lindbergh* (New York: G. P. Putnam & Sons, 1928) offers an interesting view of Lindbergh during his 1927 flight around the United States at the behest of the Daniel Guggenheim Fund.

Again, there are numerous pilot accounts from the 1920s and 1930s that give the feeling of what it was like to fly in what some writers have referred to as "aviation's Golden Age." Three of the best are: Frank Courtney, *The Eighth Sea* (Garden City, N.Y.: Doubleday & Co, 1972), Alan Wheeler, *Flying between the Wars* (Oxfordshire: G. T. Foulis & Co, 1972), and Sir Gordon Taylor, *The Sky Beyond* (Boston: Houghton Mifflin Co, 1963). Other autobiographies treat the development of aviation technology and airlines, as do a few selected biographies. Especially recommended are: F. R. "Rod" Banks, *I Kept No Diary* (Shrewsbury, England: Airlife Publications, 1978), which offers much useful information on engine development; Juan de la Cierva and Don Rose, *Wings of Tomorrow: The Story of the Autogiro* (New York: Brewer, Warren & Putnam, 1931), about the predecessor of the helicopter; Nevil Shute Norway, *Slide Rule: The Autobiography of an Engineer* (New York: William Morrow & Co, 1954), an impressive memoir by an excellent aeronautical engineer best known for his novels written under the pen name Nevil Shute; J. E. Morpurgo, *Barnes Wallis: A Biography* (New York: St. Martin's Press, 1972), about a legendary British aircraft designer and engineer; Ralph A. O'Neill, *A Dream of Eagles* (Boston: Houghton Mifflin Co, 1973), a memoir of an early airline entrepreneur; Robert Daley, *An American Saga: Juan Trippe and His Pan Am Empire* (New York: Random House, 1980), about the founder of Pan Am Airlines, balancing Matthew Josephson's earlier *Empire of the Air: Juan Trippe and the Struggle for World Airways* (New York: Harcourt, Brace & Co, 1943), reprinted (New York: Arno Press, 1972); Dean Smith, *By the Seat of My Pants* (Boston: Little, Brown, and Company, 1961), a memoir of the early days of flying the mail; Lowell Thomas and Edward Jablonski, *Doolittle: A Biography* (Garden City, N.Y.: Doubleday & Co, 1976), the best biographical study of pioneer aviator and test pilot James H. "Jimmy" Doolittle; and, finally, Igor Sikorsky, *The Story of the Winged-S* (New York: Dodd, Mead & Co, 1958), which details this Russian-American pioneer's work with flying boats. Sikorsky's work in the 1920s and 1930s, as well as his subsequent development of the helicopter, is examined in detail in Frank Delear, *Igor Sikorsky: His Three Careers in Aviation* (New York: Dodd, Mead & Co., 1976). Warren R. Young, *The Helicopters* (Alexandria, Va.: Time-Life Books, 1982) is also an informative account of the "chopper's" development and subsequent service.

Significantly, the 1920s were a time in which the profession of aeronautical engineering advanced rapidly. Critical to this development were the activities of the Daniel Guggenheim Fund for the Promotion of Aeronautics (1926–1930), a remarkable philanthropic activity that endowed schools of aeronautical engineering across the United States, established aeronautical research facilities dealing with safety in aviation and so-called "blind" flying, and helped create a "Model Air Line" run by Western Air Express between Los Angeles and San Francisco. Guggenheim activities have been examined by Richard P. Hallion in *Legacy of Flight: The Guggenheim Contribution to American Aviation* (Seattle: University of Washington Press, 1977). Theodore von Karman's memoir, written with Lee Edson, *The Wind and Beyond: Theodore von Karman, Pioneer in Aviation and Pathfinder in Space* (Boston: Little, Brown, and Co, 1967), provides an interesting personal account of one key individual's role in advancing aerospace engineering and education during the early years of aviation. Von Karman, the foremost aeronautical scientist of his time, immigrated to the United States to escape the increasingly anti-intellectual climate of Germany on the eve of Hitler's rise to power, and became director of the Guggenheim Aeronautical Laboratory at the California Institute of Technology. The transfer of the European laboratory tradition in aeronautical research and the role von Karman played are the subjects of Paul A. Hanle's *Bringing Aerodynamics to America* (Cambridge: MIT Press, 1982).

The 1920s and 1930s were marked by the emergence of practical air transportation. By the mid-1930s the first airplane capable of making a profit simply by flying passengers, the Douglas DC–3, was in large- scale production. The emergence of air transportation was inextricably linked to such other developments as air regulation and the growth of the air mail service. Ronald E. G. Davies, *A History of the World's Airlines* (New York: AMS Press, 1983) is a useful introduction. W. David Lewis and Wesley P. Newton, *Delta: The History of an Airline* (Athens, Ga.: University of Georgia Press, 1979) is a thorough and incisive examination of how one small crop-dusting company blossomed into one of America's most successful and influential air carriers. Robert J. Serling, a popular writer on airline and air transportation matters, has prepared several generally useful airline histories, especially *From the Captain to the Colonel: An Informal History of Eastern Airlines* (New York: Dial Press, 1980), and *The Only Way to Fly: The Story of Western Airlines* (Garden City, N.Y.: Doubleday, 1976). William M. Leary, *The Dragon's Wings* (Athens, Ga.: University of Georgia Press, 1976) is an excellent monographic study on the evolution of civil aviation in China. Wesley P. Newton, *The Perilous Sky: U.S. Aviation Diplomacy and Latin America, 1919–1931* (Miami: University of Miami Press, 1978), is a definitive examination of civil aviation expansion into Latin America and the often complex arrangements that had to be worked out. Henry Ladd Smith, *Airways Abroad: The Story of American World Air Routes* (Madison: University of Wisconsin Press, 1950), and Carl Solberg, *Conquest of the Skies: A History of Commercial Aviation in America* (Boston: Little, Brown, and Co, 1979) are excellent references as well as introductory guides. George E. Hopkins, *The Airline Pilots: A Study in Elite Unionization* (Cambridge: Harvard University Press, 1971) is a unique work that concentrates on the creation of the Air Line Pilots Association. Finally, Ernest K. Gann, *Fate is the Hunter* (New York: Simon & Schuster, 1961) is an excellent and evocative memoir of flying in the early days of American civil aviation by a well-known airman and novelist. Jean Potter, *The Flying North* (New York: The Macmillan Co, 1947) details the experiences of bush flying in Alaska in the 1930s.

Peter W. Brooks' classic study, *The Modern Airliner: Its Origins and Development* (London: Putnam, 1961) is still the best survey of the growth of air transport technology and the driving factors of airline operating economics. The historical series of the Federal Aviation Administration is a particularly valuable source of information on American civil aviation. Pertinent volumes include: Nick A. Komons, *Bonfires to Beacons* (Washington: Federal Aviation Administration, 1978); John R. M. Wilson, *Turbulence Aloft* (Washington: Federal Aviation Administration, 1979); Stuart I. Rochester, *Takeoff at Mid-Century* (Washington: Federal Aviation Administration, 1976); and Richard J. Kent Jr., *Safe, Separated, and Soaring* (Washington: Federal Aviation Administration, 1980). The best single-volume work is Donald Whitnah's excellent and insightful *Safer Skyways: Federal Control of Aviation, 1926–66* (Iowa City, Iowa: Iowa University Press, 1966). The impact of the growing science of meteorology was crucial to early commercial aviation, and Whitnah's one-volume *A History of the U.S. Weather Bureau* (Urbana, Ill.: University of Illinois Press, 1961) traces the important links that developed between the meteorological and aviation communities, links which mutually stimulated the development of both. Henry Ladd Smith, *Airways: The History of Commercial Aviation in the United States* (New York: Knopf, 1942) is a dated but still very useful study. Ronald E. Miller and David Sawers, *The Technical Development of Modern Aviation* (New York: Praeger, 1970) is a work that concentrates heavily on the growth of commercial aircraft technology and the economics of operating passenger aircraft.

Equally useful for understanding the technological climate from which the transport aircraft of the 1920s and 1930s sprang is Thomas Foxworth, *The Speed Seekers* (New York: Doubleday & Co, 1976) the definitive history of air racing during the 1920s and 1930s. Elsbeth E. Freudenthal, *The Aviation Business: From Kitty Hawk to Wall Street* (New York: Vanguard, 1940) is an informative and highly critical study of the growth of the aircraft manufacturing and air transport industry. Monte Duane Wright, *Most Probable Position: A History of Aerial Navigation to 1941* (Lawrence, Kans.: University Press of Kansas, 1972) is an in-depth study of the development of air navigation instruments and techniques. It was, of course, from the air mail era that commercial aviation blossomed, and, interestingly, the air mail period is one that has received too little detailed attention. Aside from Jackson's Time-Life book mentioned earlier, a useful single-volume treatment is Donald B. Holmes, *Airmail: An Illustrated History, 1793–1981* (New York: Clarkson & Potter, 1981), which is heavily illustrated and of special value to anyone interested in philately. Both supplant Page Schamburger's earlier and now dated *Tracks Across the Sky: The Story of the Pioneers of the U.S. Air Mail* (Philadelphia: Lippincott, 1964). Harold Mansfield's, *Vision: A Saga of the Sky* (New York: Duell, Sloan and Pearce, 1956) amply chronicles the early history of Boeing, developers of the first "modern" transport—the Model 247 of 1933. Finally, two works specifically examine the greatest and most influential of all airliners, the Douglas DC–3: Douglas J. Ingell's, exhaustive *The Plane That Changed The World: A Biography of the DC–3* (Fallbrook, Calif.: Aero Publishers, 1966), and Carroll V. Glines' and Wendell F. Moseley's, more popularly written, *The DC–3: The Story of a Fabulous Airplane* (Philadelphia: Lippincott, 1966).

The 1920s and 1930s were a time during which the gas- filled dirigible airship reigned supreme until a series of weather-related accidents, culminating in the *Hindenburg* holocaust, shattered forever the vision of long-range freight and passenger-carrying airships. The romance of the airship, a holdover from previous fascination with the balloon, has led to these craft being the subject of numerous popular works. Aside from Tom Crouch's new book mentioned previously, the best sources of airship history are: Douglas H. Robinson, *Giants in the Sky: A History of the Rigid Airship* (Seattle: University of Washington Press, 1973); Richard K. Smith, *The Airships Akron and Macon* (Annapolis: U.S. Naval Institute Press, 1965); Basil Collier, *The Airship: A History* (New York: G. P. Putnam's Sons, 1974); John Toland, *Ships in the Sky* (New York: Holt, 1957); Robin Higham, *The British Rigid Airship, 1908–1931* (London: G. T. Foulis & Co, 1961); Joseph Gordon Vaeth, *Graf Zeppelin: The Adventures of an Aerial Globetrotter* (New York: Harper & Bros, 1958); and James Leasor, *The Millionth Chance: The Story of the R–101* (London: Hamish Hamilton, 1957).

During the 1930s, the pace of aeronautical technology accelerated rapidly. This challenged such areas as aerospace medicine, propulsion systems, and safety during high altitude flight. Douglas Robinson, *The Dangerous Sky: A History of Aviation Medicine* (Seattle: University of Washington Press, 1973) is an excellent survey of the development of a very important field by an author who is himself a flight physician. Robert Schlaifer's exhaustive and lengthy *Development of Aircraft Engines* (Boston: Harvard University Press, 1950) and S. D. Heron's *Development of Aviation Fuels* (Boston: Harvard University Press, 1950) remain the standard reference on aero propulsion; readers desiring a briefer and less extensive introduction should consult C. Fayette Taylor, *Aircraft Propulsion* (Washington: Smithsonian Institution Press, 1971), part of the Smithsonian *Annals of Flight* series. Another volume in this series is by Stanley R. Mohler and Bobby H. Johnson, *Wiley Post, his Winnie May, and the World's First Pressure Suit* (Washington: Smithsonian Institution Press, 1971), a thorough and informative account of the activities of a noted pioneer long-distance aviator and his important research on the problems of high-altitude flight.

The development of military aviation in the 1920s and 1930s was marked by the emergence of doctrines emphasizing long-range strategic air power, tactical aviation, and naval aviation. The seeds for the employment of air power in the Second World War were planted and nurtured during the 1920s. Two useful introductions to the military aviation issues of the interwar years are Nevin's Time-Life book *Architects of Air Power* cited earlier, and Sir Arthur Hezlet's excellent survey of naval aviation, *Aircraft and Sea Power* (New York: Stein and Day, 1970). The evolution of the aircraft carrier as a capital ship has been examined in detail by Charles M. Melhorn in *Two Block Fox: The Rise of the Aircraft Carrier, 1911–1929* (Annapolis: U.S. Naval Institute Press, 1974). The major event affecting public perceptions of air power in the United States during the 1920s was the court-martial of Billy Mitchell, an outspoken and often incautiously intemperate airpower zealot. Mitchell has been the subject of numerous articles and works, but Alfred F. Hurley, *Billy Mitchell: Crusader for Air Power* (New York: Franklin Watts, 1964), new edition (Bloomington, Ind.: Indiana University Press, 1975) remains the most reliable source for information on this controversial figure. Mitchell was surrounded by a coterie of bright, ambitious, and dedicated junior officers who did much to advance the cause of a separate and independent air arm that eventually came to fruition with the creation of the United States Air Force in 1947. The contribution of one pioneer, not particularly friendly to Mitchell, is covered in John F. Shiner's penetrating *Foulois and the U.S. Army Air Corps, 1931–1935* (Washington: Office of Air Force History, 1983). The interwar years and many of these men are the subject of a major study by DeWitt S. Copp, *A Few Great Captains: The Men and Events that Shaped the Development of U.S. Air Power* (Garden City, N.Y.: Doubleday, 1980) sponsored by the Air Force Historical Foundation.

The aircraft of the interwar years have been the subject of numerous studies and monographs. The best general introductory source for material on specific aircraft types is Macmillan's *Pocket Encyclopedia of World Aircraft in Color* series, including Kenneth G. Munson's *Airliners Between the Wars 1919–1939* (New York: Macmillan, 1972), *Fighters Between the Wars 1919–1939* (New York: Macmillan, 1970), *Bombers Between the Wars 1919–1939* (New York: Macmillan, 1970), and *Flying Boats and Seaplanes Since 1910*, (New York: Macmillan, 1971). In addition, Lennart Ege, *Balloons and Airships 1783–1973*, (London: Macmillan, 1973) and Lord Ventry and Eugene Kolesnik, *Jane's Pocket Book of Airships* (New York: Collier Books, 1977), are useful pocket guides to airships.

The Ascendency of the Propeller-Driven Airplane, 1936–1945

From the late 1930s through 1945, aviation underwent many profound changes, in part because of the demands of wartime activities and requirements. Chief among these changes were the turbojet revolution, which reached its fulfillment after 1945, and the conscious recognition that the world would never quite be the same as a result of blending advanced long-range aircraft with the capabilities of modern military weapons, especially the atomic bomb. By 1945, propeller-driven long-range aircraft were flying across the world's oceans to deliver cargo and to attack targets. Yet, such was the pace of aeronautical development that their technology was already obsolescent in the face of the turbojet revolution. Truly, by the end of 1945, the dominance of the propeller-driven airplane had already begun to decline; by 1958, with the arrival of the first practical intercontinental jet transports, it would be totally swept away from the mainstream of long-range air operations.

World War II proved a tremendous "forcing function" in the evolution of military and civil aviation. In many respects, however, this was more evolutionary than revolutionary, in that the scope of wartime activities had more or less been determined by the respective development of aviation in the major combatant nations during the 1920s and 1930s.

Literally thousands of volumes have been written on the impact of air power upon the Second World War. R. J. Overy, *The Air War, 1939–1945* (New York: Stein and Day, 1980) offers excellent coverage and insights, including a revealing examination on aircraft production. The Higham and Hezlett works cited previously examine facets of the employment of air power during the war, and are the best introductory references to the subject. Additionally, there has been a sizable body of official histories dealing either exclusively or in part with the war in the air. The standard American reference on the war in the air is Wesley Frank Craven and James Lea Cate, eds, *The Army Air Forces in World War II*, 7 Vols (Chicago: University of Chicago Press, 1948–1958), reprint edition (Washington: Office of Air Force History, 1984). The U.S. Navy and Marine Corps story is told by Samuel Eliot Morison, *History of United States Naval Operations in World War II*, 15 vols, (Boston: Little, Brown and Company, 1947–1962), and Robert L. Sherrod, *History of United States Marine Corps Aviation in World War Two* (San Rafael, Calif.: Presidio Press, 1980). Haywood S. Hansell, Jr., *The Air Plan that Defeated Hitler* (Atlanta: Printed by Higgins-McArthur/Logino & Porter, Inc., 1972), reprinted (New York: Arno Press, 1980) is part memoir, part analytical history by one of the great Air Force planners; the study covers prewar events as well as planning and strategic operations in the war against Germany.

LITERATURE

One very significant study of the military aircraft procurement process is I. B. Holley, *Buying Aircraft: Materiel Procurement for the Army Air Forces (United States Army in World War II, Special Studies)* (Washington: Office of the Chief of Military History, Department of the Army, 1964), part of the official *United States Army in World War II* series. Benjamin S. Kelsey, *The Dragon's Teeth* (Washington: Smithsonian Institution Press, 1982) is an important memoir and historical analysis of the problems affecting military procurement in America during the 1920s and 1930s. Clark G. Reynolds, *The Fast Carriers: The Forging of an Air Navy* (New York: McGraw-Hill Book Co, 1968) is a brilliant, incisive history of American carrier operations during the Second World War. Readers desiring a less exhaustive, more popular, and heavily illustrated introductory work are referred to Reynolds' *The Carrier War* (Alexandria, Va.: Time-Life Books, 1982) a volume in the previously mentioned Time-Life *Epic of Flight* series. A. D. Turnbull and C. L. Lord, *History of U.S. Naval Aviation* (New Haven: Yale University Press, 1949) contains much useful information, and Theodore Roscoe, *On the Seas and In the Skies: A History of the U.S. Navy's Air Power* (New York: Hawthorn Books, 1970) is also a lively account.

U.S. Army Air Forces' operations have been the subject of numerous popular works. However, they are often characterized by a "gee whiz" approach to the subject that obscures, or even supplants, more serious interpretations and discussions. Readers thus should choose very carefully among available sources, being especially cautious when dealing with books written immediately after the war and even into the 1950s and early 1960s. One major study that furnishes useful insights on the U.S. Army Air Forces is by DeWitt S. Copp, *Forged in Fire: Strategy and Decisions in the Air War Over Europe, 1940–1945* (Garden City, N.Y.: Doubleday & Co, 1982).

Nothing was as symbolic of the role air power played in the Second World War as the massive Anglo-American bombing raids launched against Axis targets. There is a wealth of literature on this subject, but again, readers have to choose with caution. Noble Frankland, *The Bombing Offensive Against Germany* (London: Faber & Faber, 1965) is an excellent introduction. Lee B. Kennett, *A History of Strategic Bombing* (New York: Charles Scribner's Sons, 1982) is a concise, scholarly study that places the World War II experience in the broad perspective of previous efforts to harness aviation to war. Max Hasting's *Bomber Command* (New York: The Dial Press, 1979) offers a clear and often disturbing insight into the evolution of British bombing policy and operational techniques. Roger A. Freeman, *The Mighty Eighth* (Garden City, N.Y.: Doubleday & Co, 1970) is an encyclopedic history of the Eighth Air Force, the U.S. Army Air Forces' principal striking arm in the war against Nazi Germany. Two of the key air offensives of the war were raids on Schweinfurt, Germany, a center of ball-bearing production, and on the Ploesti, Rumania oilfields. Martin Caidin, *Black Thursday* (New York: Dutton, 1960) is an excellent account of the planning and execution of the costly Schweinfurt strike, and Ploesti has been the subject of two excellent works: Leon Wolff, *Low-Level Mission* (Garden City, N.Y.: Doubleday & Co, 1957), and the definitive *Ploesti: The Great Ground-Air Battle of 1 August 1943* (New York: Random House, 1962) by James Dugan and Carroll Stewart. James M. Merrill, *Target Tokyo: The Halsey-Doolittle Raid* (Chicago: Rand McNally, 1964) provides a good account of this interesting joint air-naval strike against the Japanese mainland in April 1942. Carroll V. Glines, *Doolittle's Tokyo Raiders* (Princeton, N.J.: D. Van Nostrand, 1964) and his later, *Four Came Home* (Princeton, N.J.: D. Van Nostrand, 1966), provide the recollections and impressions of those who actually flew on the raid, including those who were imprisoned by the Japanese. Curtis E. LeMay and MacKinlay Kantor, *Mission With LeMay: My Story* (Garden City, N.Y.: Doubleday & Co, 1965) is an interesting memoir of a major figure in American airpower, and has much useful information on LeMay's role in both European and Pacific strategic bomber operations. Allen Andrews, *The Air Marshals: The Air War in Western Europe* (New York: W. W. Morrow and Co, 1970) is a good introduction to the personalities who were responsible for both Allied and Axis conduct of the air war. A book which focuses on American operations and the role of one key individual is Thomas Coffey, *Hap: The Story of the U.S. Air Force and the Man Who Built It: General Henry H. "Hap" Arnold* (New York: The Viking Press, 1982). David MacIsaac's *Strategic Bombing in World War II: The Story of the United States Strategic Bombing Survey* (New York: Garland Publishing, 1976) recounts how the survey was organized and accomplished. Finally, Fletcher Knebel and Charles W. Bailey, *No High Ground* (New York: Harper, 1960) is an excellent account of the first atomic bomb strikes at Hiroshima and Nagasaki.

The war in the Pacific, while not as extensively treated as the war in Europe, has, nevertheless, been the subject of a number of works. George C. Kenney, *General Kenney Reports: A Personal History of the Pacific War* (New York: Duell, Sloan and Pearce, 1949) covers the Southwest Pacific Theater with thoroughness and insight. Edwards Park, *Nanette* (New York: W. W. Norton, 1977) is an evocative and highly personal account of a fighter pilot's combat tour in New Guinea, and ranks among the finest pilot memoirs of the war. Charles A. Lindbergh, *The Wartime Journals of Charles A. Lindbergh* (New York: Harcourt Brace Jovanovich, 1970) is a major source for this controversial airman's views and interpretations of wartime events, including excellent sections on his wartime service in the Southwest Pacific. Claire Lee Chennault, *Way of a Fighter: The Memoirs of Claire Lee Chennault* (New York: Putnam, 1949), edited by Robert Hotz, is a useful memoir on the famed leader of the "Flying Tiger" volunteers who flew for China. Thomas G. Miller, *The Cactus Air Force* (New York: Harper and Row, 1969) is an excellent account of Allied air operations over the Solomon Islands in 1942–1943. The Japanese side of the air war has been well treated by surviving airmen and military historians. Especially recommended are the following: Masatake Okumiya and Jrio Horikoshi with Martin Caidin, *Zero* (New York: Dutton, 1956); Saburo Sakai with Martin Caidin and Fred Saito, *Samurai* (New York: Dutton, 1957), a pilot memoir by Japan's top-scoring surviving fighter ace; Mitsuo Fuchida and Masatake Okumiya, *Midway: The Battle that Doomed Japan* (Annapolis: U.S. Naval Institute Press, 1955); and Rikihei Inoguchi and Tadashi Nakajima, with Roger Pineau, *The Divine Wind: Japan's Kamikaze Force in World War II* (Westport, Conn: Greenwood Press, 1958), an insightful account of the origins and employment of the *Kamikaze* against Allied fleets. Finally, Wilbur H. Morrison, *Point of No Return: The Story of the 20th Air Force* (New York: Times Books, 1979) has a popularly written and reliable account of the bombing offensive that destroyed Japan.

One of the critical periods of the war was the Battle of Britain in 1940, which resulted in a clear British victory and the thwarting of Hitler's plans for an invasion of the British Isles. The literature and pilot memoirs on both sides of this conflict are extensive. Derek Wood and Derek Dempster, *The Narrow Margin* (London: Hutchinson, 1961) is the standard historical account. Peter Townsend's later, *Duel of Eagles* (New York: Simon and Schuster, 1970) is a masterful blend of memoir and history. Francis K. Mason's encyclopedic *Battle Over Britain* (Garden City, N.Y.: Doubleday & Co, 1970) is the best source for statistical data, as well as giving a blow-by-blow daily account of the struggle. The depth of Mason's research is remarkable.

The various air forces of the Second World War combatants have been treated in detail, especially that of Nazi Germany. Three useful introductory works on the combatant air forces are volumes in the Time-Life series: Ralph Barker, *The RAF at War* (Alexandria, Va.: Time-Life Books, 1981); Time-Life eds, *The Luftwaffe* (Alexandria, Va.: Time-Life Books, 1982); and Russell Miller, *The Soviet Air Force* (Alexandria, Va.: Time-Life Books, 1983). The Soviet Air Force has already been the subject of several popular works. Alexander Boyd, *The Soviet Air Force since 1918* (New York: Stein and Day, 1977) is an excellent survey history, as is Kenneth R. Whiting, *Soviet Air Power, 1917–1978* (Maxwell AFB: Air University Press, 1979); Ray Wagner and Leland Fetzer, *The Soviet Air Force in World World II: The Official History* (Garden City, N.Y.: Doubleday & Co, 1973) is a faithful translation of the Soviet official history published by the Ministry of Defense, Moscow, and thus contains many of the quaint quasi-judgmental vagaries of the original. Von Hardesty, *Red Phoenix: The Rise of Soviet Air Power, 1941–45* (Washington: Smithsonian Institution Press, 1982) is a definitive history of Soviet Air Force operations during the Second World War.

Denis Richard's and Hilary St. George Saunders' three-volume *The Royal Air Force, 1939–1945*, rev ed (London: Her Majesty's Stationary Office, 1974–1975) is the general official British history. British airmen have written dozens, if not hundreds, of memoirs which, for the most part, are well-written and highly informative. The strategic war waged by the RAF is covered in the excellent official history by Sir Charles K. Webster and Noble Frankland, *The Strategic Air Offensive Against Germany, 1939–1945,* 4 Vols (London: Her Majesty's Stationary Office, 1961).

The Luftwaffe has continued to fascinate popular writers and historians through the years. Two "insider" accounts that avoid many of the self-serving generalizations used in other memoirs of the Nazi era are: Adolf Galland, *The First and the Last* (London: Fontana, 1970) by the former chief of German fighter forces; and Werner Baumbach, *The Life and Death of the Luftwaffe* (New York: Coward-McCann, 1960), written by the former chief of Germany's Bomber Command. David Irving, *The Rise and Fall of the Luftwaffe: The Life of Field Marshal Erhard Milch* (Boston: Little, Brown & Co, 1974) is a provocative, absorbing, and insightful examination of the Luftwaffe from the perspective of its chief architect. Finally, W. H. Tantum, IV and E. J. Hoffschmidt, *The Rise and Fall of the Luftwaffe 1933–1945,* (Old Greenwich, Conn.: WE, Inc, 1969) provides an excellent summary history of that ineptly led and confused organization. Smaller air forces have not received their fair due, though Jerzy Cynk, *History of the Polish Air Force, 1918–1968* (Reading, U.K.: Osprey Publishing, 1972) is an exception, and a study that could serve as a model for others to follow.

One of the truly significant aspects of aviation from 1936 through 1945 was the development of global air transport networks. While the Allied nations had a keen awareness of the importance of air transport, the Axis nations did not; Germany, for example, ignored military air transportation, with the loss of Stalingrad offering clear evidence of this critical weakness in Nazi strategic doctrine. Some good works exist on Allied air transport operations. Oliver LaFarge, *The Eagle in the Egg* (Boston: Houghton, 1949) is a popular and reliable history of the U.S. Army Air Forces Air Transport Command written shortly after the war. William H. Tunner, the wartime manager of America's famed "Hump" aerial supply route to China, has written an excellent memoir of his work and the lessons learned from the Hump (lessons subsequently applied during the Berlin Airlift after the war), *Over the Hump* (New York: Duell, Sloan and Pearce, 1964).

Various works have been written on specific aspects of aeronautics during the Second World War. David Irving, *The Mare's Nest: The German Secret Weapons Campaign and British Countermeasures* (London: William Kimber, 1964) is an excellent study of the development of German "robot" weapons, the V–1 "buzz bomb" cruise missile and the V–2 ballistic missile. R.V. Jones, *The Wizard War: British Scientific Intelligence, 1939–1945* (New York: Coward, McCann and Geoghegan, 1978) while a memoir is the most complete description to date of the scientific duel in the air war between the British and the Germans. It is likely this book will become a classic. Brian Johnson, *The Secret War* (New York: Methuen, 1978) is a good introduction to the wartime scientific research of both the Allies and Axis, most of which was related to aviation. Leslie Simon, *German Research in World War II, An Analysis of the Conduct of Research* (New York: J. Wiley & Sons, 1947) is the standard reference on Nazi Germany's scientific war, and James Phinney Baxter, III, *Scientists Against Time* (Boston: Little, Brown and Co, 1946), reprinted (Boston: MIT Press, 1968) is an excellent study of the workings of the wartime Office of Scientific Research and Development, which conducted numerous aviation-related programs. Constance Babington-Smith, *Evidence in Camera: The Story of Photographic Intelligence in World War II* (London: Chatto and Windus, 1958) is a remarkably complete survey of wartime work by an individual who was herself a distinguished practitioner. Glenn Infield, *Unarmed and Unafraid* (New York: Macmillan Co, 1970) is an anecdotal survey history of aerial reconnaissance, with much useful material on the Second World War.

A potpourri of other works have been published. Perry McCoy Smith, *The Air Force Plans for Peace, 1943–1945* (Baltimore, Md.: Johns Hopkins University Press, 1970) is an excellent administrative history of how the Air Force began adapting for its postwar roles even as fighting was still underway. Sally Van Wagenen Keil, *Those Wonderful Women in their Flying Machines* (New York: Rawson & Wade, 1979) is a witty and informative history of the wartime Women's Air Force Service Pilots (WASPs) program. Alan Osur, *Blacks in the Army Air Forces During World War II* (Washington: Office of Air Force History, 1977) is an excellent official study of how the U.S. Army Air Forces utilized black troops and flight crews, often confronting problems of basic prejudice in the process. A good popular source is Charles E. Francis, *The Tuskegee Airmen: The Story of the Negro in the U.S. Air Force* (Boston: Bruce Humphries, 1955). Richard C. Lukas, *Eagles East: The Army Air Forces and the Soviet Union, 1941–1945* (Tallahassee, Fla.: Florida State University Press, 1970) is a thorough study of the important aerial lend-lease activity between the United States and the Soviet Union.

Readers interested in the work of aircraft designers during this time period will find a surprising amount of literature on the subject. The following works are highly recommended, and reflect the international character of aviation: Edward H. Heinemann and Rosario Rausa, *Ed Heinemann: Combat Aircraft Designer* (Annapolis: U.S. Naval Institute Press, 1980), a memoir by the designer of a series of important American attack airplanes; Jrio Horikoshi, *Eagles of Mitsubishi: The Story of the Zero Fighter* (Seattle: University of Washington Press, 1981), by the designer of Japan's famed wartime fighter; Alexander Lippisch, *The Delta Wing: History and Development* (Ames, Iowa: Iowa State University Press, 1981), by a noted German aircraft designer; Heinz Conradis, *Design for Flight: The Kurt Tank Story* (London: MacDonald, 1960), about the designer of Germany's Focke-Wulf aircraft; Ernst Heinkel, *Stormy Life* (New York: Dutton, 1956), a memoir about the German aircraft industry in the interwar years; and Alexander Yakovlev, *Notes of an Aircraft Designer* (New York: Arno Press, 1972), which sheds some useful light on Soviet wartime design. Two other important works that involve British wartime aeronautical and defense research are Ronald W. Clark's biography, *Tizard* (Cambridge: MIT Press, 1965) and C. P. Snow's well-known *Science and Government* (Cambridge: Harvard University Press, 1961). Both are insightful commentaries on the conduct of science and technology in wartime

Great Britain, and both offer a detailed perspective on the famous conflict between Sir Henry Tizard and Lord Cherwell.

As with other aspects of aviation history, the aircraft of the combatant powers have received coverage bordering on the repetitive and excessive. Kenneth Munson, *Bombers, Patrol, and Transport Aircraft, 1939–1945* (New York: Macmillan, 1969) and *Fighters, Attack, and Training Aircraft, 1939–1945* (New York: Macmillan, 1969) are reliable and informative. William Green is a well-known author of definitive reference works on World War II, and two of his best are the republished *Famous Fighters of the Second World War* (Garden City, N.Y.: Doubleday & Co, 1975) and *Famous Bombers of the Second World War* (Garden City, N.Y.: Doubleday & Co, 1976). Many other more specialized works, most of which deal with a single aircraft or "family" of aircraft also exist but are beyond the scope of this introduction to the field.

The Turbojet Revolution and the Supersonic Breakthrough, 1945–1957

Between 1945 and 1957, the scope of aviation changed radically. During this time period, civil long-range aviation became the dominant form of domestic and international passenger transportation, surpassing the accumulated accomplishments of a century of railroad and motor vessel transportation. Flight speeds quadrupled due to the development of refined high-speed aerodynamics and advanced propulsion by jet and rocket engines. The turbojet revolution permitted developing highly efficient high-speed aircraft; the attainment of flight faster than sound—supersonic flight—opened up a whole new range of speeds and altitudes to be explored.

The major technical development influencing the course of postwar aviation was the appearance of the gas turbine (turbojet) engine. It was largely the product of a few farsighted inventors who worked outside the mainstream of contemporary propulsion research and development. Only later was gas turbine technology seized upon by a propulsion industry previously totally committed to the piston engine and propeller. The early jet era is one that has not been as thoroughly examined as it should be. Nevertheless, a number of useful books are available for researchers. Schlaifer and Heron's previously cited *Development of Aircraft Engines and Fuels* is excellent for the development of the jet engine in Britain, Germany, and the United States. Edward Constant, *The Origins of the Turbojet Revolution* (Baltimore: Johns Hopkins University Press, 1980) is a detailed examination of the actual evolutionary process that led to the jet engine. Sir Frank Whittle, *Jet: The Story of a Pioneer* (New York: Philosophical Library, 1954) is an autobiography which tells the trials and tribulations of how "The Father of the Jet Engine" went about winning support and actually designing and building the then radical powerplant. John Grierson's rare but fascinating *Jet Flight* (London: Sampson Low, Marston & Co, Ltd, 1946) combines history and autobiography to trace the story of the jet engine and important wartime development work. Also Walter J. Boyne and Donald S. Lopez, *The Jet Age: Forty Years of Jet Aviation* (Washington: Smithsonian Institution Press, 1979) is a useful introductory reader, as is E. T. Wooldridge, *Jet Aviation: Threshold to a New Era* (Washington: Smithsonian Institution Press, 1981). Grover Heiman's anecdotal *Jet Pioneers* (New York: Duell, Sloan and Pearce, 1963) is a good, readable popular reference. William Green and Roy Cross, *The Jet Aircraft of the World* (Garden City, N.Y.: Hanover House, 1955) remains the best single reference work on the wide range of experimental aircraft built during the first fifteen years of jet flight.

The supersonic breakthrough constituted the most important development in aeronautics since the time of the Wrights' first flights, yet it has only been the subject of two books. Richard P. Hallion, *Supersonic Flight: Breaking the Sound Barrier and Beyond* (New York: Macmillan Co, 1972) discusses the problem of supersonic flight and how a specialized research aircraft program was established to confront it. Charles Burnet, *Three Centuries to Concorde* (London: Mechanical Engineering Publications

Ltd, 1979) is equally thorough in examining British efforts to achieve flight faster than sound. The individuals most responsible for ensuring that reliable data on supersonic flight would be acquired were a group of highly skilled test and research pilots.

Test flying and flight research have often been the subject of misinterpretation and sensationalism. Only two books have been written that attempt to analyze the role of test pilots and their contributions to aerospace development: the previously cited *Testing Time* by Constance Babington-Smith, and Richard P. Hallion, *Test Pilots: The Frontiersmen of Flight* (Garden City, N.Y.: Doubleday & Co, 1981). In addition, an institutional history of a NASA Research Center, Richard P. Hallion's, *On the Frontier: Flight Research at Dryden 1946–1981* (Washington: NASA, 1984) reveals the complexity of making technological choices.

The test pilots who flew these early supersonic experimental aircraft have left a number of interesting memoirs, most of the "as told to" variety. The following can all be recommended: Roland Beamont, *Testing Years* (London: Ian Allan, 1980); William Bridgeman and Jacqueline Hazard, *The Lonely Sky* (New York: Henry Holt & Co, 1955); Eric Brown, *Wings on My Sleeve* (Shrewsbury, England: Airlife Publications, 1978); A. Scott Crossfield and Clay Blair, *Always Another Dawn* (Cleveland: World Publishing Co, 1960); Neville Duke and Alan W. Mitchell, *Test Pilot* (London: A. Wingate, 1953); Jay Miller, *The X–Planes: X–1 to X–29* (Marine-on-St. Croix: Specialty Press, 1983); Frank K. Everest and John Guenther, *The Fastest Man Alive* (New York: Dutton, 1958); James J. Haggerty, *First of the Spacemen* (New York: Duell, Sloan and Pearce, 1960), a biography of Iven C. Kincheloe; Tony Le Vier and John Guenther, *Pilot* (New York: Harper & Row, 1954); Mike Lithgow, *Mach One* (New York: Allan Wingate, 1954); William Lundgren, *Across the High Frontier* (New York: Morrow, 1955), a biography of Charles E. Yeager; and Peter Twiss, *Faster Than the Sun* (London: MacDonald, 1963). Tom Wolfe's readable *The Right Stuff* (New York: Farrar, Straus & Giroux, 1979) is humorous and enjoyable, but flawed by errors of fact and interpretation. Though not strictly about flight testing, Jacqueline Cochran, *The Stars at Noon* (Boston: Little, Brown, 1954) is an interesting memoir by the first American woman to fly faster than sound. Finally, while written for a young adult audience, Don Dwiggins, *Flying the Frontiers of Space* (New York: Dodd, Mead & Co, 1982) is a useful introductory overview of flight research from the first supersonic flights to the Space Shuttle.

An excellent survey of air power in the post-war years is *Air Power in the Nuclear Age* by M.J. Armitage and R.A. Mason (Urbana: University of Illinois Press, 1983). General William W. Momyer discusses American events in *Airpower in Three Wars* (Washington: Department of the Air Force, 1978). One of the first major demonstrations of air power in that era was the Berlin Airlift, where Allied air transports resupplied the city of Berlin in the face of a Soviet blockade. Walter P. Davison, *The Berlin Blockade: A Study in Cold War Politics* (Princeton: Princeton University Press, 1958) is the definitive treatment of this subject. The "Cold War" quickly turned hot in Asia, as war broke out in Korea and Indochina. Robert Frank Futrell, *The United States Air Force in Korea, 1950–1953* (New York: Duell, Sloan and Pearce, 1961), revised edition (Washington: Office of Air Force History, 1983), is a brilliant synthesis of politics, technology, and combat history. Malcolm W. Cagle and Frank A. Manson, *The Sea War in Korea* (Annapolis: U.S. Naval Institute Press, 1957) is equally excellent on the naval aviation side. Bernard Fall, *Hell in a Very Small Place: The Siege of Dien Bien Phu* (Philadelphia: Lippincott, 1967) is a thorough account of how not to run a military campaign, and how military air power proved incapable of breaking an increasing stranglehold. (It is interesting to compare this work with Bernard Nalty's account of the siege of Khe Sanh cited in the next section; at Khe Sanh, air power proved decisive).

The Air Force itself underwent a variety of technological, administrative, strategic, and tactical changes during the period from 1945 through 1957. Alfred Goldberg's previously cited *A History of the United States Air Force* offers a reliable account of key developments in this time period as does such a popular study as Carroll V. Glines, *The Compact History of the United States Air Force* (New York: Hawthorn Books, 1963). Robert Frank Futrell, *Ideas, Concepts, Doctrine: A History of Basic Thinking in the United States Air Force, 1907–1964,* (Maxwell AFB, Ala.: Air University, 1971) is an indispensible reference for anyone following the administrative and doctrinal history of the Air Force and American air power. Herman Wolk, *Organizing the Postwar Air Force, 1943–1947* (Washington: Office of Air Force History, 1984) explains the emergence of the independent service. Thomas A. Sturm, *The USAF Scientific Advisory Board: Its First Twenty Years, 1944–1964* (Washington: USAF Historical Division Liaison Office, 1967) is an informative tracing of efforts to integrate scientific advising into the Air Force planning process. It complements Nick A. Komons, *Science and the Air Force: A History of the Air Force Office of Scientific Research* (Arlington, Va.: Office of Aerospace Research, 1966). The breadth of both of these fine books takes them into the middle 1960s. Alan L. Gropman, *The Air Force Integrates, 1945–1964* (Washington: Office of Air Force History, 1978) is an excellent account of how the Air Force came to grips with the desegregation and integration drive of the 1950s and 1960s. During this time, the aircraft flown by the Air Force were also changing rapidly; Marcelle Size Knaack's *Post World War II Fighters, 1945–1973* (Washington: Office of Air Force History, 1978), the first of the multivolume *Encyclopedia of U.S. Air Force Aircraft and Missile Systems* being prepared by the Office of Air Force History, is an excellent and authoritative guide to a class of aircraft that were the first to exploit the lessons learned from the turbojet revolution and the supersonic breakthrough.

As with other aspects of aviation history, the aircraft of the 1945–1957 time period have been the subject of a variety of popular works. Again, the best introductory source works are: Kenneth G. Munson's, *Airliners Since 1946* (New York: Macmillan, 1972), and *Private Aircraft, Business and General Purpose, Since 1946* (New York: Macmillan, 1967); John R. and Michael J. H. Taylor's, *Jane's Pocket Book of Commercial Transport Aircraft* (New York: Macmillan, 1974), *Jane's Pocket Book of Military Transport and Training Aircraft* (New York: Macmillan, 1974), and *Jane's*

Pocket Book of Major Combat Aircraft (New York: Macmillan, 1973). An excellent guide to Soviet aircraft of the postwar period is William Green and Gordon Swanborough, *The Observer's Soviet Aircraft Directory* (New York: Frederick Warne & Co, Inc, 1975).

True vertical flight via helicopters was first achieved shortly before the Second World War, but it was not until the postwar years that the helicopter came of age, especially for search and rescue operations. The history of the helicopter is another subject that deserves greater attention; helicopter-like toys had appeared as early as the Renaissance. H. F. Gregory, *Anything a Horse Can Do: The Story of the Helicopter* (New York: Reynal & Hitchcock, 1944) is an entertaining and informative history of helicopters during the formative years of vertical flight. Paul Lambermont and Anthony Pirie, *Helicopters & Autogyros of the World,* rev ed (New York: A. S. Barnes & Co, 1970) is the definitive reference guide to rotorcraft through the years, though Kenneth G. Munson, *Helicopters and Other Rotorcraft Since 1907* (New York: Macmillan, 1969) is a colorful and useful introductory work.

The worldwide aerospace industry underwent profound changes in the years after 1945, taking on characteristics both multinational and conglomerate. Oddly, a single good reference work on the growth of the aerospace industry remains to be written, though some attempts have resulted in generally useful references. The following offer useful information on American and foreign aerospace industry developments: William G. Cunningham, *The Aircraft Industry: A Study in Industrial Location* (Los Angeles: Morrison, 1951); John B. Rae, *Climb to Greatness: The American Aircraft Industry, 1920–1960* (Cambridge: MIT Press, 1968); Charles D. Bright, *The Jet Makers: The Aerospace Industry: An Anthology* (Cambridge: MIT Press, 1968); Almarin Phillips, *Technology and Market Structure: A Study of the Aircraft Industry* (Lexington, Mass.: D.C. Heath & Co, 1971); E. T. Wooldridge, *Winged Wonders: The Story of the Flying Wings* (Washington: Smithsonian Institute Press, 1983); and Herman O. Stekler, *The Structure and Performance of the Aerospace Industry* (Berkeley: University of California Press, 1965). Derek Wood's *Project*

Cancelled (Indianapolis: Bobbs-Merrill, 1975), while suffering from a sometimes sensationalist approach, is a generally useful account of the near collapse of Britain's aircraft industry from mismanagement and misdirection in the late 1940s and 1950s.

One interesting and controversial aspect of post-1945 aerospace research was the employment of many scientists and engineers who had previously served the Third Reich. Clarence Lasby, *Project Paperclip* (New York: Athenaeum, 1971) traces these individuals and how they came to work for the United States. Alexander Yakovlev, *Fifty Years of Soviet Aircraft Construction* (Washington: NASA/National Science Foundation/ Israel Program for Scientific Translations, 1970), while often bland, is nevertheless informative on the emergence of the Soviet aircraft industry. A more useful but less official tracing of the Soviet aircraft industry is Heinz J. Nowarra and G. R. Duval, *Russian Civil and Military Aircraft, 1884–1969* (London: Fountain Press, 1971). Two books prepared during this time period are particularly valuable: Jerome Hunsaker, *Aeronautics at Mid-Century* (New Haven: Yale University Press, 1952), by the chairman of the National Advisory Committee for Aeronautics, the predecessor of today's NASA, and William F. Ogburn, *The Social Effects of Aviation* (New York: Houghton Mifflin, 1946), the latter one of the first major attempts at technological and social forecasting.

The Space Age and the Maturation of Aeronautics, 1957–1982

The launching of *Sputnik I* in October 1957 had a profound impact upon American society. As a result of this event, American education was totally revamped, the NACA was replaced by the NASA, and a "space race" began between the United States and the Soviet Union. Competition and rivalry in science and technology was accompanied by big power rivalry for influence among the Third World nations. Traditional animosities flared into war in the Middle East and Southeast Asia. Aviation and aerospace development played crucial roles in all of these conflicts.

LITERATURE

Aeronautics since 1957 has not been the subject of too many substantial works; rather, developments in this time period have been examined in periodical literature. Three of the best periodical sources for information on aviation since 1957 have been McGraw-Hill's journal *Aviation Week and Space Technology* (New York: McGraw-Hill, weekly); *Astronautics and Aeronautics,* the professional journal of The American Institute of Aeronautics and Astronautics (New York: American Institute of Aeronautics and Astronautics, monthly); and *Flight International* (London: IPC Transport Press, Ltd, weekly), good for the European perspective. Many of the major aerospace development programs of the 1960s and 1970s were marked by controversy.

Aside from a welter of sensationalist works written from staunchly partisan viewpoints, there has been a dearth of really incisive studies on this time period. Three very useful works, treating respectively civilian and military aeronautics, are: Kenneth Owen, *Concorde: New Shape in the Sky* (London: Jane's Publishing Co, 1982), about the controversial Anglo-French Supersonic Transport Aircraft (SST); Mel Horwitch, *Clipped Wings: The American SST Conflict* (Cambridge: MIT Press, 1982); and Robert F. Coulam, *Illusions of Choice: Robert McNamara, The F–111, and the Problem of Weapons Acqusition Reform* (Princeton: Princeton University Press, 1977), a painstaking work which replaces Robert J. Art's earlier and flawed *The TFX Decision: McNamara and the Military* (Boston: Little, Brown, 1968).

The controversy over such programs as the F–111 and the C–5 highlight the problems of weapons acquisition in an era of expensive high technology requiring lengthy periods of development. J. Ronald Fox, *Arming America: How the U.S. Buys Weapons* (Cambridge: Harvard University Press, 1974), and Jacques S. Gansler, *The Defense Industry* (Cambridge: MIT Press, 1980) furnish thorough analyses of the aerospace industry and national defense procurement. Merton J. Peck and Frederick M. Scherer, *The Weapons Acquisition Process* (Cambridge: Harvard University Press, 1962) offers a definitive and highly detailed examination of the actual procedural stages of weapons evolution. Thomas A. Marschak, *The Role of Project Histories in the Study of R & D* (Santa Monica, Calif.: The RAND Corporation, 1965) is a useful examination of selected key defense programs involving missiles, aircraft, engines, and radar systems and the lessons that their development histories offer to defense planners. Arnold S. Levine, *Managing NASA in the Apollo Era* (Washington: NASA, 1982) is an excellent examination of the management of large-scale, time-critical technological programs, and the complexities of running such programs in an intensive political environment. Two complementary studies use early ballistic missile development to draw large lessons about how the United States procures weapons systems. Michael H. Armacost, *The Politics of Weapons Innovation: The Thor-Jupiter Controversy* (New York: Columbia University Press, 1969) examines early medium-range missile development and the interservice problems of competition and collaboration. Edmund Beard, *Developing the ICBM: A Study in Bureaucratic Policies* (New York: Columbia University Press, 1976) concentrates on the early years of intercontinental ballistic missile development and the evolution of postwar Air Force missile doctrine.

The following three books offer occasionally sharp criticism of the defense acquisition process and the spread of advanced weapons technology: Robert Gilpin, *American Scientists and Nuclear Policy* (Princeton: Princeton University Press, 1962); Ralph E. Lapp, *Arms Beyond Doubt: The Tyranny of Weapons Technology* (New York: Cowles Book Co, 1970); and Ralph E. Lapp, *The Weapons Culture* (New York: W. W. Norton, 1968). Bernard Brodie, *Strategy in the Missile Age* (Princeton: Princeton University Press, 1959), while dated, offers useful insights into the relationship between doctrine, changing technology, and the complexities of weapons choice. Robert S. McNamara's provocative *The Essence of Security: Reflections in Office* (New York: Harper & Row, 1968) is a good introduction to the views of this highly controversial Secretary of Defense at a critical time in American military affairs. Finally, Norman R. Augustine, *Augustine's Laws and Major System Development Programs* (New York: American Institute of Aeronautics and Astronautics, 1982) is an essential reference to the often perplexing problems confronting defense planners. Written in a humorous and ironic style in the form of "laws," this book, by a former undersecretary of the Army and president of a major aerospace corporation, is a provocative and insightful analysis of the problems of weapons system procurement and the management of research and development.

The protracted war in Southeast Asia spawned volumes of popular studies of the bombing campaigns, air fighting, aircraft, and personalities involved. Much of this was of a "buff" nature, particularly the great body of literature concerning individual aircraft types. Nevertheless, as with other aspects of the conflict, there have been some useful studies produced, although the Vietnam era is clearly one that requires a well-integrated military, social, and political history volume of the kind the late Bernard B. Fall so eloquently produced on France's earlier experience in Indochina.

The Office of Air Force History currently has a multivolume series, *The United States Air Force in Southeast Asia* in preparation, the first volume of which, Robert F. Futrell and Martin Blumenson, *The Advisory Years to 1965* (Washington: Office of Air Force History, 1981), has established the style, tone, and theme for the remainder of the series. An earlier, and essentially photographic history of Air Force involvement, Carl Berger, *et al, The United States Air Force In Southeast Asia, 1961–1973* (Washington: Office of Air Force History, 1977), is a useful introduction to the subject. Earl H. Tilford, *Search and Rescue in Southeast Asia, 1961–1975* (Washington: Office of Air Force History, 1980), covers the all-important effort to rescue downed aircrew and other personnel from hostile territory. Roger P. Fox, *Air Base Defense in the Republic of Vietnam, 1961–1973* (Washington: Office of Air Force History, 1979) is almost a manual for the subject as well as a history of the U.S. experience. Bernard C. Nalty, *Air Power and the Fight for Khe Sanh* (Washington: Office of Air Force History, 1973), is an excellent analysis of a critical seige in which air power proved decisively important. William A. Buckingham, *Operation Ranch Hand: The Air Force and Herbicides in Southeast Asia, 1961–1971* (Washington: Office of Air Force History, 1982), is an analysis of this often controversial program. Jack S. Ballard, *Development and Employment of Fixed-Wing Gunships, 1962–1972* (Washington: Office of Air Force History, 1982), discusses the evolution of one of the war's most distinctive and innovative weapons. Ray L. Bowers, *Tactical Airlift* (Washington: Office of Air Force History, 1983) covers the Southeast Asia war comprehensively in that crucial mission area.

The USAF Air War College Airpower Research Institute, *USAF Southeast Asia Monograph Series* offers more specialized and detailed analysis of specific campaigns, activities, and problems. Of special value are the following works: A. J. C. Lavelle, ed, *The Tale of Two Bridges* and *The Battle for the Skies Over North Vietnam,* Vols 1 & 2, *USAF Southeast Asia Monograph Series* (Washington: Government Printing Office, n.d.); A. J. C. Lavelle, ed, *The Vietnamese Air Force, 1951–1975, An Analysis of its Role in Combat* and *Fourteen Hours at Koh Trang,* Vol 3, *USAF Southeast Asia Monograph Series* (Washington: Government Printing Office, n.d.); A. J. C. Lavelle, ed, *Last Flight From Saigon,* Vol 4, *USAF Southeast Asia Monograph Series* (Washington: Government Printing Office, n.d.); Alan L. Gropman, *Airpower and the Airlift Evacuation of Kham Duc,* Vol 5, *USAF Southeast Asia Monograph Series* (Maxwell AFB, Ala.: Airpower Research Institute, 1979); James R. McCarthy, George B. Allison, and Robert E. Rayfield, *Linebacker II: A View from the Rock,* Vol 6, *USAF Southeast Asia Monograph Series* (Maxwell AFB, Ala.: Airpower Research Institute, 1979); and, finally, Donald K. Schneider, *Air Force Heroes in Vietnam,* Vol 7, *USAF Southeast Asia Monograph Series* (Maxwell AFB, Ala.: Airpower Research Institute, 1979). Two other specialized studies deserve mention: John J. Tolson's *Airmobility, 1961–1971 (U.S. Army Vietnam Studies)* (Washington: Department of the Army, 1973); and Robert Frank Futrell, *et al, Aces and Aerial Victories: The United States Air Force in Southeast Asia, 1965–1973* (Washington: Office of Air Force History, 1976), on the air combat over North Vietnam from 1965 through 1973.

One of the more significant technological developments of the Vietnam era was the emergence and employment of precision-guided "smart" weaponry, and the use of unmanned remotely piloted vehicle (RPV) technology to undertake a variety of hazardous missions at minimal cost and risk to human lives. As a result of the Vietnam experience, and subsequent lessons learned from other conflicts, the RPV occupies a secure place as a tool in the military arsenals of many nations. Arthur Reed, *Brassey's Unmanned Aircraft* (London: Brassey's Publishers Ltd, 1979) is a well-written reference on the development, current status, and projected future uses of military remotely piloted vehicles, popularly but often mistakenly called "drones." William Wagner, *Lightning Bugs and Other Reconnaissance Drones* (Fallbrook, Calif.: Aero Publishers, Inc, 1982) is a generally useful and anecdotal account of the Ryan Corporation's highly successful Firebee family of RPVs and their use in Southeast Asia, but it is marred by a lack of sources and a narrow internalist perspective. John W. R. Taylor and Kenneth G. Munson, *Jane's Pocket Book of Remotely Piloted Vehicles* (New York: Collier Books, 1977) is an excellent reference to the various RPVs that have been developed, as well as a useful guide to their early history. Finally, Roger A. Beaumont, "Rapiers Versus Clubs: The Fitful History of 'Smart Bombs,'" *Journal of the Royal United Services Institute* 126, No. 3 (Sep 1981) is an excellent reference for the history of precision guided weaponry and its influence on tactics and doctrine.

The war in Southeast Asia and various small "brushfire" conflicts have also been the subject of some popular studies, memoirs, and the like. One very useful reference is the Senator Mike Gravel edition of the so-called "Pentagon Papers", first illegally released by antiwar activist Daniel Ellsberg, entitled *The Pentagon Papers: The Defense Department History of United States Decisionmaking on Vietnam* (Boston: Beacon Press, 1971), a five-volume work. Jack Broughton, *Thud Ridge* (Philadelphia: Lippincott, 1969) is a lively and excellent memoir by a combat pilot who is bitter over American bombing policy in the early years of the air war "up North." Robinson Risner, *The Passing of the Night: My Seven Years as a Prisoner of the North Vietnamese* (New York: Random House, 1974) is an unforgettable and moving memoir of a POW's struggle to survive. Benjamin Schemmer, *The Raid* (New York: Harper, 1976) is a thorough account of the ill-fated attempt to rescue POWs believed held in the Son Tay prison camp outside Hanoi. Finally, Bernard C. Nalty, George M. Watson, and Jacob Neufeld, *An Illustrated Guide to the Air War Over Vietnam: Aircraft of the Southeast Asia Conflict* (New York: Arco/Salamander, 1981) is a useful and reliable guide to the equipment flown by both sides of that long war. Sterling Seagrave, *Soldiers of Fortune* (Alexandria, Va.: Time-Life Books, 1981) is an informative and lively account of how aerial mercenaries have influenced small wars through the years. Christopher Robbins, *Air America* (New York: G. P. Putnam's Sons, 1979) offers the reader a good behind-the-scenes look at a little-known government-sponsored air transport operation.

Unrelated to warfare and mercenary flying, but having many of the same "roughing it" aspects, is Arctic bush flying in Alaska and Canada. This is a subject rich in anecdotes and worthy of detailed treatment; Harmon Helmericks, *The Last of the Bush Pilots* (New York: Knopf, 1970) is a good introduction.

To the public, the greatest expression of aeronautics and astronautics in the 1960s and 1970s was found in the space programs of the United States and the Soviet Union. While not vast, a respectable body of literature concerned with manned and unmanned spaceflights already exists. Very few of the influential early theoreticians and pioneers have been the subject of biographies. A notable exception is Robert Goddard, an American physicist who launched the drive towards space with his firing of the world's first liquid fuel rocket in 1926. Milton Lehman, *This High Man* (New York: Farrar, Straus, 1963) is a thoughtful and sympathetic biography of Goddard which examines the occasionally mystical and secretive nature of this somewhat tragic figure. Goddard's own papers are available in the three-volume *The Papers of Robert H. Goddard* (New York: McGraw-Hill, 1970), edited by his widow Esther C. Goddard and rocket pioneer G. Edward Pendray. For other sources on the early days of rocketry, interested readers are advised to consult the following: Wernher von Braun and Frederick I. Ordway, *History of Rocketry and Space Travel,* 3rd rev ed (New York: Crowell, 1975), which features an excellent bibliography;

LITERATURE

David Baker, *The Rocket: The History and Development of Rocket & Missile Technology* (New York: Crown, 1978); Kenneth Gatland, *et al, The Illustrated Encyclopedia of Space Technology* (New York: Harmony, 1981); Eugene M. Emme, *The History of Rocket Technology: Essays on Research, Development, and Utility* (Detroit: Wayne State University Press, 1964), an excellent introduction to the "hard technology" side of rocket-propelled space research; and, finally, Willy Ley, *Rockets, Missiles, and Men in Space* (New York: The Viking Press, 1968), which is still quite useful despite the fact that it first appeared in 1944.

The growth of manned spaceflight reached such major proportions during the 1960s and 1970s that, understandably, social scientists attempted to place its development in the context of other developments in the history of technology. This led to a notable series of essays edited by Bruce Mazlish, *The Railroad and the Space Program: An Exploration in Historical Analogy* (Cambridge: MIT Press, 1965).

The international aspects of space—involving the rights of nations, the conduct of international affairs, and joint exploration and utilization of space—are receiving increasing attention. A useful introductory study on this important subject is Arnold W. Frutkin's *International Cooperation in Space* (Englewood Cliffs, N.J.: Prentice-Hall, 1965), an examination of the various considerations that can influence the conduct of technology and science on a global scale. George S. Robinson, *Living in Outer Space* (Washington: Public Affairs Press, 1975) offers the legal perspective on spaceflight.

The history of rocketry can be arranged to reflect four major periods: the early years through mid-1957; Sputnik and its aftermath through the first utilization of space; the "heroic era" of manned spaceflight through the landing of Apollo 11 on the moon; and, finally, the post-Apollo years.

The early years of rocketry were most notable for the work of Goddard in America, and the German rocketeers in Nazi Germany. The best work on

Nazi Germany's rocketry efforts is Frederick I. Ordway and Mitchell R. Sharpe, *The Rocket Team* (New York: Thomas Y. Crowell, 1979). A slightly different but no less valuable perspective is that of David Irving's aforementioned *The Mare's Nest*. R. F. Pocock, *German Guided Missiles of the Second World War* (New York: Arco, 1967) is a well written technical reference to the specific weapon systems developed at Peenemunde and other German test sites.

Early postwar American work is also discussed in Ordway and Sharpe, *The Rocket Team*. Milton Rosen, *The Viking Rocket Story* (New York: Harper, 1955) is a classic and gracefully written saga of the first indigenous American upper atmospheric sounding rocket. The first American satellite effort, the Vanguard project, is thoroughly examined by Constance McLaughlin Green and Milton Lomask, *Vanguard: A History* (Washington: Smithsonian Institution Press, 1971), including the shattering impact Sputnik had upon American science and the Vanguard effort. James R. Killian, Dwight D. Eisenhower's "Missile Czar," has amply captured the turbulence of the post-Sputnik era in his memoir *Sputnik, Scientists, and Eisenhower: A Memoir of the First Special Assistant to the President for Science and Technology* (Cambridge: MIT Press, 1977). For years the late Charles S. Sheldon of the Library of Congress maintained meticulous records on the Soviet space program. His *United States and Soviet Progress in Space* (Washington: Library of Congress, 1978) is a good overview of Soviet efforts. Two good popular accounts are Nicholas Daniloff, *The Kremlin and the Cosmos* (New York: Knopf, 1972) and James E. Oberg's more recent and contemporary *Red Star in Orbit* (New York: Random House, 1981). One of the aspects of spaceflight has been remote sensing for defense and space science purposes. Philip J. Klass, *Secret Sentries in Space* (New York: Random House, 1971) while dated, is a good popular account of the development of reconnaissance satellites and their impact upon international affairs. Merton E. Davies and Bruce C. Murray, *The View From Space: Photographic Exploration of the Planets* (New York: Columbia University Press, 1971) and Frederick I. Ordway, *Pictorial Guide to Planet Earth* (New York: Crowell, 1975) are good references on what has been accomplished in terms of peaceful applications of remote sensing. Leonard Jaffe, *Communications in Space* (New York: Holt, Rinehart and Winston, 1966), while dated, is a useful introduction to the technology and history of communications satellite technology. A very different—and valuable—kind of work is Alfred Bester's *The Life and Death of a Satellite* (Boston: Little, Brown, 1966) which treats the whole problem of how key individuals work together—sometimes at the expense of their health—to develop a successful space program.

The Science Policy Research Division of the Library of Congress, Congressional Research Service has issued an excellent and exhaustive reference on the entire United States civilian space effort since 1958, *United States Civilian Space Programs, 1958–1978* (Washington: Government Printing Office, 1981), the first volume of a projected series of reports prepared for the Subcommittee on Space Science and Applications of the U.S. House of Representatives' Committee on Science and Technology. Finally, Ken Hechler, *Toward the Endless Frontier: History of the Committee on Science and Technology 1959–1979* (Washington: U.S. Congress, House of Representatives, 1980), offers the perspective of space decision-making from a long-standing member of this critical House Committee.

The "heroic era" of manned spaceflight has been admirably treated by a series of NASA-sponsored histories that are primarily project-oriented, but which also include social, political, and economic factors. Prior to the onset of the Space Shuttle Program, the American manned spaceflight effort involved Mercury, Gemini, Apollo, Skylab, and the Apollo-Soyuz test project, the latter a joint United States-Soviet Union mission. The following works are especially recommended to the interested reader: Lloyd S. Swenson, James M. Grimwood, and Charles C. Alexander, *This New Ocean: A History of Project Mercury* (Washington: NASA, 1966); Barton C. Hacker and James M. Grimwood, *On the Shoulders of Titans: A History of Project Gemini* (Washington: NASA, 1977); R. Cargill Hall, *Lunar Impact: A History of Project Ranger* (Washington: NASA, 1981); Courtney G. Brooks, James M. Grimwood, and Lloyd S. Swenson, *Chariots for Apollo: A History of Manned Lunar Spacecraft* (Washington:

NASA, 1979); and Edward C. Ezell and Linda N. Ezell, *The Partnership: A History of the Apollo-Soyuz Test Project* (Washington: NASA, 1978).

John Logsdon's *The Decision to Go to the Moon* (Cambridge: MIT Press, 1970) constitutes an insightful and important reference on the political environment surrounding the decision to undertake the Apollo lunar landing effort. A handy reference and introduction to the Apollo program and its social, political, technological, and scientific significance is Richard P. Hallion and Tom D. Crouch, eds, *Apollo: Ten Years Since Tranquillity Base* (Washington: Smithsonian Institution Press, 1979), a series of essays by authorities in various fields ranging from rocket technology to space art and lunar geology. Also useful are two works: Kerry Joels, *Apollo to the Moon: A Dream of Centuries* (Washington: Smithsonian Institution Press, 1982), prepared as a Smithsonian exhibit gallery guide, and "The Moon Landing and Its Aftermath," printed in a special issue of the *Michigan Quarterly Review,* 18, No. 2 (Spring 1979). Henry S. F. Cooper, *13: The Flight That Failed* (New York: Dial Press, 1973) is a gripping account of the Apollo mission that nearly failed disastrously and tragically in space, but which was rescued by creative decisionmaking and professional excellence. The scientific harvest gleaned from Apollo and especially the Apollo-Soyuz mission is the subject of Farouk El-Baz's *Astronaut Observations from the Apollo-Soyuz Mission* (Washington: Smithsonian Institution Press, 1977), which offers the reader a good insight into the potentiality of remote sensing. Two noted artists, H. Lester Cooke and James D. Dean, who were administrators of NASA's art program in which leading artists were invited to record their impressions of the space program, have collected a reflective and stimulating visual record of the American space effort in *Eyewitness to Space: Paintings and Drawings Related to the Apollo Mission to the Moon* (New York: Abrams, 1971).

Spaceflight has so far produced few good commentaries, but three are recommended. Norman Mailer, *Of a Fire on the Moon* (Boston: Little, Brown, 1970) discusses what Apollo meant to Mailer and the so-called "Aquarius Generation." Tom Wolfe, *The Right Stuff,* previously cited, examines the world of the test pilot and astronaut, and the occasional tensions therein. Michael Collins, *Carrying the Fire: An Astronaut's Journeys* (New York: Farrar Straus Giroux, 1974) is a humorous, thoughtful, and lively recollection of the Gemini and Apollo programs and a host of other things, by the former command module pilot of Apollo 11. It is among the finest aviation memoirs written to date.

There is, of course, a very large body of literature that may be termed "space futurism." Much of this speculation falls between factual extrapolation from today's technology and outright science fiction. One glimpse that promises to be a landmark book in the literature of space utilization is physicist Gerard K. O'Neill's *The High Frontier: Human Colonies in Space* (New York: William Morrow, 1977) which postulates a future of high technology and cost-effective space colonies orbiting the earth, a vision that is hotly debated by technologists, scientists, enthusiasts, and social commentators.

Several surveys and reference works concerned with spaceflight should be mentioned. The Congressional Research Service of the Library of Congress issued *Astronauts and Cosmonauts: Biographical and Statistical Data,* rev ed, (Washington: Government Printing Office, 1978), which is the single best source of biographical information on Soviet, Soviet bloc, and American space explorers. Lynne C. Murphy, *Rockets, Missiles, and Spacecraft of the National Air and Space Museum* (Washington: Smithsonian Institution, Press, 1976) is a useful introductory guide to the space holdings of the Smithsonian Institution. Kenneth Gatland, *Manned Spacecraft* (New York: Macmillan, 1976) and *Missiles and Rockets* (New York: Macmillan, 1975) are useful guides to the hardware of the space age, as are Reginald Turnill's *The Observer's Book of Unmanned Spaceflight* (New York: Frederick Warne and Co, Inc, 1974) and *The Observer's Book of Manned Spaceflight* (New York: Frederick Warne and Co, 1975). J. C. D. Blaine, *The End of an Era in Space Exploration: From International Rivalry to International Cooperation* (San Diego, Calif.: American Astronautical Society, 1976) and Frederick C. Durant, a series of edited essays, *Between Sputnik and the Shuttle: New Perspectives on American Astronautics* (San Diego, Calif.: American Astronautical Society, 1981), are volumes in the historical series of the American Astronautical Society, offering tentative interpretations on what the space program has meant to the United States and the Soviet Union. Paul A. Hanle and Von Del Chamberlin, *Space Science Comes of Age* (Washington: Smithsonian Institution Press, 1981), a series of edited essays, offers a quick perspective of the scientific impact of unmanned space exploration. Clearly, however, the field of space exploration is one that will continue to require incisive study and research.

Index

INDEX

INDEX

INDEX

INDEX

INDEX

INDEX

INDEX